THE
PHYSICS
OF
STAR
WARS

THE
PHYSICS
OF
STAR
WARS

The Science Behind a Galaxy Far, Far Away

PATRICK JOHNSON, PHD

ADAMS MEDIA

NEW YORK   LONDON   TORONTO   SYDNEY   NEW DELHI

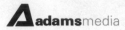

Adams Media
An Imprint of Simon & Schuster, Inc.
57 Littlefield Street
Avon, Massachusetts 02322

First Adams Media trade paperback edition NOVEMBER 2017

ADAMS MEDIA and colophon are trademarks of Simon and Schuster.

For information about special discounts for bulk purchases, please contact Simon & Schuster Special Sales at 1-866-506-1949 or business@simonandschuster.com.

The Simon & Schuster Speakers Bureau can bring authors to your live event. For more information or to book an event contact the Simon & Schuster Speakers Bureau at 1-866-248-3049 or visit our website at www.simonspeakers.com.

Interior design by Colleen Cunningham
Interior image © 123RF/Jiří Tomek

Manufactured in the United States of America

10 9 8 7 6 5 4 3 2

Library of Congress Cataloging-in-Publication Data
Johnson, Patrick, author.
The physics of Star Wars / Patrick Johnson, PhD.
Avon, Massachusetts: Adams Media, 2017.
Includes index.
LCCN 2017026420 (print) | LCCN 2017037390 (ebook) | ISBN 9781507203309 (pb) | ISBN 9781507203316 (ebook)
LCSH: Physics--Miscellanea. | Star Wars films--Miscellanea. | Science in motion pictures.
LCC QC75 (ebook) | LCC QC75 .J64 2017 (print) | DDC 530--dc23
LC record available at https://lccn.loc.gov/2017026420

ISBN 978-1-5072-0330-9
ISBN 978-1-5072-0331-6 (ebook)

CONTENTS

PLANET-BASED TRANSPORTATION 65

SPACE TRAVEL 85

HANDHELD WEAPONRY 109

HEAVY WEAPONRY 135

THE FORCE 161

ROBOTICS 185

OTHER TECH 209

ACKNOWLEDGMENTS

This book would not have been possible without the support system surrounding me. Primary thanks need to be given to my partner, Susan Tyler, for her extensive advising on matters of both grammar and Star Wars. She is a beautiful genius, and I cannot thank her enough for the effort she has put into this book. Additional thanks need to be given to my friends and family who provided insightful conversations about all matters in this book. They helped check my math, my writing, and my sanity.

INTRODUCTION

Thank you for picking up *The Physics of Star Wars*! This book uses the fantastic world of Star Wars to explore physics in a new way. If you are already a Star Wars fan, you know that the stories take place in a galaxy far, far away, so the laws of physics should still apply. On the other hand, these are obviously works of fiction; is there any point in applying those laws? This book makes the case that it is both fun and worthwhile to do so. Sometimes the physics shown in the movies is spot on while on other occasions it would require advanced technology or new discoveries in the realm of physics to make sense. Either way, science is about the critical thinking process needed to tackle a problem rather than the specific situation in which the problem appears. There's no reason we can't consider Yoda force-lifting rocks instead of pulleys lifting blocks!

To that end, this book contains a series of categorized topics. Each topic is given a brief introduction and backstory followed by analysis of the physics in the Star Wars universe as well as where current technology or scientific understanding puts us relative to it. These topics draw only on what is seen in the movies; there would just be too much to talk about otherwise.

Still, the films don't always provide all the answers needed to explain a physics topic. What exactly is a lightsaber? Is it a plasma or a beam of light? Depending on the source you consult outside of the movies, it could be either. In this book, what is depicted in the movies is taken as definitely true, but other canon sources are considered when needed. If there is not a clear conclusion that can be drawn from what is shown in the film, several possible theories are discussed.

Obviously, the movies do not provide technical specifications for all the ships and planets. When those numbers were needed, relative measurements from freeze-frames or timed length of scenes were used. For the sake of clarity, not all calculations are shown in full detail. If you want to reproduce them on your own, you can do that with the help of an introductory physics book. The beauty of science is that no matter who or where you are, you should be able to reproduce the results of another person's work.

I hope this book makes both Star Wars and physics more exciting for you!

SPACE

ANATOMY OF THE GALAXY

WHEN All films	
WHERE The galaxy	
CHARACTERS All	
PHYSICS CONCEPTS Gravity, dark matter, circular motion	

SHORT INTRODUCTION/BACKGROUND

Our universe is big—really big. We live on planet Earth, inside of one solar system located on one arm of the Milky Way galaxy. The Milky Way is one of many galaxies in what is called a galactic supercluster, of which there are many (and that's just in the observable universe!). Here are some numbers to give you a sense of how big all of this is. Our solar system is 0.00096 light-years across; the Milky Way is about 100,000 light-years across; its local supercluster is about 500,000,000 light-years across. There are estimated to be about ten million superclusters in the observable universe. How did these all form? How do galaxies get their shapes?

BACKSTORY

Star Wars is primarily set in a region of space referred to as "the galaxy." Images of this region show it as a spiral galaxy (much like the Milky Way) divided into regions like the core, inner rim, outer rim, etc. The galactic core is home to the governing center of Coruscant, whereas the outer rim has planets such as Tatooine, run by gangsters and outlaws. Characters frequently refer to various "systems" or "star systems" (rather than solar systems) in the galaxy, then use a planet name when identifying a specific system (e.g., the Hoth system). This would be like calling our solar system the Earth system rather than the Sun system. This makes sense if most travel is to/from one planet within a system, but it might make the galaxy seem smaller than it really is. Does the number of planets/systems identified in Star Wars make sense for one galaxy? Is the layout of these systems realistic?

THE PHYSICS OF STAR WARS

In *The Force Awakens*, R2-D2 projects a map of the entire Star Wars galaxy, a spiral galaxy with two arms. In general spiral galaxies consist of a dense cluster of stars around a central point (most often a supermassive black hole), plus a number of less dense regions extending outward like arms. R2's map is consistent with that description, but we don't know for certain just how densely packed some of the systems are in the core versus the outer rim.

Since we don't know the literal distances involved, let's analyze this from the perspective of governments and power. The influence of the Galactic Empire is felt differently across the galaxy. Galaxies are often hundreds of thousands of light-years across, which means there are great distances separating the various star systems. This would explain how Yoda lived undiscovered on Dagobah or how the rebels hid on a moon of Yavin. This also explains how a faction of the Empire could secretly create a massive weapon capable of destroying planets.

DIFFERENT ORBITAL RATES

Star systems close to the core of the galaxy would orbit the supermassive black hole rather quickly. Since outer stars orbit at different rates than inner stars (much like planets in our solar system rotate at different speeds), a planet in a core star system could be on the same side or opposite side of the galaxy as a planet in an outer star system at various points. So, depending on what year it is, the distance between Naboo and Coruscant could vary by around 50 percent, given their relative distances from the galactic center. Further, core worlds would experience mild time dilation from being close to the supermassive black hole. In a 100-year life span, someone living on Coruscant could expect to age about a few hours less than an outer-rim counterpart.

The minimal information presented in the films about other galaxies is also consistent with what we'd expect. For example, in *Attack of the Clones*, Obi-Wan is given directions to a planet called Kamino (where the clone troopers are being bred) relative to the Rishi Maze, a neighboring dwarf galaxy. Dwarf galaxies are a real thing; they orbit larger galaxies, sort of like a moon. Unlike moons, dwarf galaxies are destroyed over millions of years by this relationship with a larger body. Why does this happen?

The Moon pulls on Earth as it orbits, causing ocean tides; a similar process happens with dwarf galaxies. Roughly speaking, gravity will have a stronger pull on the near side of a dwarf galaxy than on the far side. Therefore, some stars and planets will be pulled closer as they orbit the larger galaxy. Because the closer stars are still attracted to the rest of the dwarf galaxy (and the rest of the dwarf galaxy is still attracted to them), the galaxy will stretch out. Because galaxies aren't super dense to begin with (unlike moons), this will end in their being pulled apart.

THE PHYSICS OF REAL LIFE

Like the galaxy seen in Star Wars, the Milky Way is a spiral galaxy. There aren't too many other options—the universe appears to contain primarily elliptical and spiral galaxies. Other types have been observed and there is plenty to debate about how to classify them, but rather than get bogged down in taxonomy, let's focus on the physics behind spiral galaxies.

One promising theory suggests that there are different densities of stars in spiral galaxies because of what is known as a density wave. This is akin to what happens in a traffic jam. The cars in front slow down, causing the cars behind them to slow down. The cars in front can then speed up again, but the shock created by that initial slowing can last long after the original slowing. With galaxies, it's not that stars slow down; instead, they feel a gravitational attraction to the stars around them as well as to the supermassive black hole in the center. Think of it as a complex game of tug-of-war. The

UNIVERSAL GRAVITATION

In 1687 Newton published his *Principia Mathematica*, which contains his law of universal gravitation. This states that the force of gravity is given by $F = \dfrac{Gm_1m^2}{r^2}$. At the time it had long been known that objects moving in a circular path were subject to the rules of acceleration, $\dfrac{v^2}{r}$. Using Newton's second law (the rate of change in momentum of an object is directly proportional to the force applied to it), you can relate force and acceleration to find a prediction for the speed of an orbiting object. Its speed should be given by $\sqrt{\dfrac{Gm}{r}}$. This means that the speed should decrease as you move away from the center of a galaxy.

tug from the supermassive black hole keeps the stars traveling in a circular orbit, but neighboring stars tug back which can cause a star to speed up or slow down as it orbits the center of the galaxy.

Scientific study of galaxies and their shapes has led to the proposal of the existence of dark matter. Although many observations come together to suggest its existence, the speed at which stars orbit their galactic center is an important indicator. According to Newton's law of gravity and the laws of circular motion, the speed of stars should slow down as you go farther from the galactic center. Measurements have indicated that this is not the case; stars have a near constant speed after a certain radius.

Given that the speed of stars is nearly constant, either Newton's laws (or the laws of circular motion) are wrong in this situation or we are missing an element of the equations. If Newton's laws don't apply, we may need to develop a theory of modified gravity. Alternatively, we may have misrepresented the mass in the galaxy. This is where the idea comes from that there is a large quantity of unseen matter (hence the name dark). Combining this result with a number of other observations that have been made, it seems that the only explanation for all the anomalies in measured gravity (as of now) is the existence of dark matter.

BIODIVERSITY IN THE GALAXY

"Why do I sense we've picked up another pathetic life-form?"
—Obi-Wan Kenobi (Episode I)

WHEN All films	
WHERE The galaxy	
CHARACTERS All	
PHYSICS CONCEPTS The Drake equation	

SHORT INTRODUCTION/BACKGROUND

There is a chance we will discover evidence that life once existed on other planets, but as far as we know Earth is the only planet in the universe where life currently exists. We know that there are billions of stars in the Milky Way plus billions of galaxies in the universe. Are we really alone in this universe? If we are not alone, why haven't we heard from anybody else? Are other species purposely avoiding us? Have they attempted to contact us, and we just don't know how to listen yet?

BACKSTORY

According to the beginning of every Star Wars movie, the saga takes place in a single galaxy. As the stories unfold, there is no shortage of biodiversity. Sure, the main characters are mostly human, but one glance inside the Mos Eisley cantina will show you numerous other species. Hundreds of different species appear in the Star Wars movies. Is it reasonable to assume that one galaxy would have hundreds of different alien life-forms? How many planets developed life themselves rather than simply sustaining alien colonization?

THE PHYSICS OF STAR WARS

In the Anatomy of the Galaxy section, we established that Star Wars takes place in a spiral galaxy consistent with scientific knowledge of such galaxies. In order to determine if the biodiversity seen in the movies is also consistent, we need more specifics about the galaxy than the films provide. We'll have to combine typical real-life data with reasonable assumptions based on the movies.

A typical spiral galaxy is around a few hundred thousand light-years across and has several hundred billion stars. The Milky Way, for example, is at least 100,000 light-years across and has at least one hundred billion stars. A light-year is the distance that light can travel in a year, or about six trillion miles (these numbers are often too big for us to wrap our heads around, which is why scientists use units such as light-years).

If we are generous and assume that the Star Wars galaxy has about a trillion stars (which would make it larger than any observed spiral galaxy), would we expect it to have the level of biodiversity depicted? The short answer to this is yes, because a few hundred different species are shown, and the Earth has around eight million species. However, such numbers are misleading. At no time during the films do we follow an entomologist studying all the different species of bugs on Dagobah, for instance (but if such a movie is made, you'll know where Disney got the idea). For a more accurate comparison, we might assume that the hundreds of species correspond to, say, two hundred different planets of origin, and then see if that is a high or a low prevalence of life.

Our current method for estimating life in the universe uses something called the Drake equation. Although the equation was never intended to be exact, it can give an estimated order of magnitude for the number of intelligent species in the universe (or in this case, in the Star Wars galaxy). The Drake equation is as follows: $N = R^* \cdot f_p \cdot n_e \cdot f_l \cdot f_i \cdot f_c \cdot L$, in which N is the number of species, R^* is the rate of star formation, f_p is the fraction of stars that have planets, n_e is the average number of planets per star that can support life, f_l is the fraction of Earth-like planets that have life, f_i

CONSTRUCTING THE DRAKE EQUATION

A few of the parameters for the Drake equation are known from scientific measurement or are not relevant for our particular estimate. For instance, it appears that nearly every star has planets, so most estimates put f_p close to 1. Also, R* has been measured to be somewhere around 1–3 stars per year. The average number of planets that are habitable, n_e, is currently estimated to be around 0.4. Notice how a lot of the equation's parameters are communications related? The Drake equation is technically trying to estimate the number of alien civilizations that might attempt to communicate with us. If all we care about is the number of species (and not their communication skills), we don't need to worry about the terms representing the fraction that will develop intelligent life, the fraction that will develop communications, and the length of those civilizations.

Even ignoring terms specific to communications, other parameters in this equation are hotly debated. There is evidence that shortly after conditions became favorable for life on Earth, life formed. Some scientists claim this means that the fraction of habitable planets that develop life should be close to 1. Others say that this is survivor bias; they think that the scientists claiming the number is close to 1 are using the argument that "this happened, therefore this is always what happens," and that that is a fallacy. Furthermore, if it is so likely for life to spontaneously form on a habitable planet, shouldn't it have happened more than once on Earth? All life that we know of today can be traced back to a common ancestor, indicating that life has only spontaneously formed once.

is the fraction of planets that have yielded intelligent life, f_c is the fraction of intelligent life-forms that have developed communication, and L is the lifetime of these civilizations' communications.

Since we have most of the parameters needed for the Drake equation, we can start plugging in numbers. Using the estimate of two hundred planets with life, the assumption that 100 percent of stars have planets, and 40 percent of them are habitable, and a trillion stars in the galaxy, only five out of every ten billion planets would need to form life for the biodiversity of the movies to be reasonable.

Let's remind ourselves what all these numbers mean. We have manipulated the Drake equation for use in the Star Wars galaxy such that we are solving for f_1 (the fraction of habitable planets that have produced life) because we can estimate N (the number of planets with life) by watching the movies. In real life, scientists estimate f_1 so that they can solve for N. When scientists estimate f_1, they use values anywhere from $\sim\!\frac{1}{10}$ to $\sim\!1/10^{40}$. Thus, five in every 10 billion planets is plausible, but that isn't saying much.

THE PHYSICS OF REAL LIFE

At this time we can say with certainty that there is at least one planet with moderately intelligent life. To know how much other life we might expect, we can use the Drake equation again, but keep in mind that this equation yields a huge range of values. We just don't know how often habitable planets produce any kind of life (again, it may be $\frac{1}{10}$ or $\frac{1}{10}^{40}$). The value we got from the Star Wars galaxy was actually on the lower end of that spectrum, so let's use it. Current estimates for the number of stars in our universe are around 10^{19}, so, plugging those into the Drake equation, we'd expect there to be 2 billion planets with life on them.

If there are 2 billion planets with life, it seems as if we should meet (or hear from) a new alien every week or so. Not so fast! When you compare that with trillions of planets in existence, the chances of even a second planet with life on it in our Milky Way are quite small. Even if there were a second planet in the galaxy with life on it, it would probably be tens of thousands of light-years away. When we consider interacting with life that far away, there are many complex factors; let's break them down.

Consider a hypothetical timeline for communications between us and an alien civilization; we'll call them Bothans. In an ideal scenario, the Bothans would be ready and waiting to receive signals from other life. Radio communications were invented on Earth in 1895. So, imagine in 1895 humans had sent a signal across the galaxy to the Bothan planet, and the Bothans sent one back. If the Bothans lived on a planet orbiting Alpha Centauri (the next closest star to Earth after the Sun), the response from the Bothans could have arrived on Earth around 1905 or so. Given the billions of stars in the Milky Way alone, it is much more likely the Bothans would be significantly farther away. If the Bothans were halfway between us and the center of the Milky Way, we wouldn't receive the Bothan signal until the year 51895.

That scenario assumes all conditions are ideal other than the distance between us and the Bothans. For one, the Bothans may not be prepared to receive any signal from any direction. Similarly, humans in 1895 had no reason to be aware of the Bothans or their location, so they would have had to broadcast the radio signals in all directions. Once the Bothans received the radio signal, how would they know what to do with it?

One also has to consider the part of the Drake equation (which we have been ignoring) about how long a species will last. Even though the universe has been around for more than thirteen billion years, that doesn't mean civilizations have existed that long. Even if we assume that an intelligent life-form came into existence on another planet at some point in the history of the universe, it is unlikely that this civilization existed in a time frame where it could communicate with us. Signals take time to travel from one location to another.

To be more concrete, if the Neanderthals had figured out interstellar communication, and they knew exactly where to point their message, it would be just arriving at a planet on the other side of our galaxy. If the next closest planet was in our nearest galaxy (Andromeda), then the Neanderthal signal would still have another 2.5 million years to go. Even if this communication reached a planet with life, that life might well be single-celled creatures with no way to understand the signal. Much in the way that the Neanderthals would still be waiting for their signal to arrive in Andromeda, another civilization trying to communicate with us may still be waiting for their signal to reach us. Historically, empires like the Egyptians or Romans did not last longer than 500 years. If the Greeks told the conquering Romans, "We started a dialogue with Zeus, please listen to a certain part of the sky for the response in a few thousand years," not only would the Romans need to not ignore this crazy request, they would need to make sure it was passed on for 50,000 years.

RELATIVE AGING OF CHARACTERS

WHEN Luke's and Leia's lifetimes	
WHERE Tatooine and Alderaan	
CHARACTERS Luke Skywalker, Princess Leia Organa	
PHYSICS CONCEPTS General relativity, special relativity	

SHORT INTRODUCTION/BACKGROUND

When Einstein first proposed his theory of special relativity in 1905, the scientific community did not accept his theories. Einstein suggested that the passage of time changes depending on how fast a person (or reference frame) is moving. Crazy as it sounds, if you send one half of a set of twins away in a rocket ship only to return later, she'll find that she is younger than her brother (this idea is known as the twin paradox). Since the early 1900s we have done extensive tests of the special and general theories of relativity

TIME DILATION

Not convinced about time dilation? One of the fundamental tenets of relativity is that the speed of light is a constant regardless of our speed and travels at about 186,000 miles per second. Imagine setting up two mirrors about a foot apart so that light can bounce back and forth vertically between them. The light will be able to bounce back and forth between the mirrors about 500,000,000 times every second. If we use this device to measure time, we can define one second as the time it takes for the light to travel back and forth 500,000,000 times. This will be accurate even if the room the mirrors are in is moving to the right at a constant speed (as the light moves up and down). Now imagine the room is indeed moving and someone watches it fly by. This observer will see the light move in a zigzag pattern as it bounces up and down because of the relative sideways motion. This person sees light traveling extra distance at the same speed (remember, the speed of light is a constant), so something about the time has to have changed.

and found that they are perfectly consistent. Does this mean that one twin is indeed younger when returning from a long interstellar journey?

BACKSTORY

When we are first introduced to Luke Skywalker and Princess Leia Organa, they don't appear to have any connection. The original novelization of *A New Hope* puts Luke at twenty years old (or "twice as old as a ten-year-old vaporator") while Leia is described as being eighteen years old. We later discover that Luke and Leia are actually twins born about nineteen years prior to the events of *A New Hope*. It has been rationalized that Luke's and Leia's ages were described approximately so it's not inconsistent if they are nineteen years old, but could it be that they are different ages? According to theories of relativity, could additional traveling make Leia two years younger than her twin brother?

THE PHYSICS OF STAR WARS

The "speed" of time's passage involves a process called time dilation, one of the main consequences of Einstein's theory of special relativity. Consider the situation of Luke and Leia. Luke is delivered to Tatooine as a baby and stays there until he and Obi-Wan travel toward Alderaan. Leia on the other hand is taken to Alderaan and grows up to be a diplomat. With her zooming around on spaceships at very fast speeds, time will move more slowly for her relative to the stationary Luke.

Just how much would Leia need to zoom around the galaxy such that Luke is two years older than she is when they meet up again? As discussed previously, time will move slower for the moving Leia as compared to stationary Luke. Technically, to do this calculation, we would need to know Leia and Luke's relative speeds at all times in their lives. Let us consider a few specific examples to get a sense of how possible this is.

If all Leia did was travel away from Polis Massa (the asteroid where she and Luke were born) and return at age eighteen to find Luke there aged twenty years, she would have to travel about 44 percent of the speed of light (or about 81,000 miles per second) the entire time she was away. She would have to travel even faster to account for Luke's presumed travel from Polis Massa to Tatooine plus all of her downtime on Alderaan. If we were to assume that Leia had a grand total of fifteen years of stationary life and five years of travel time, it would require her to constantly be traveling at 70 percent of the speed of light, or about 130,000 miles per second.

This does not take into account relative aging due to the effects of general relativity, though. General relativity states that just being in a gravitational field causes time to slow down. If you take into account these effects, using estimated diameters for Alderaan and

Tatooine, and assume their densities are somewhere around Earth's density, Luke's and Leia's ages would only be different by approximately one second in the span of twenty years, so it is mostly a negligible effect.

THE PHYSICS OF REAL LIFE

This all sounds pretty fantastical, but all experimentation indicates that this is how time really works. Although to date we haven't been able to develop a manned rocket fast enough to do this experiment, we have done a related experiment with twin atomic clocks. In 1971 Joseph Hafele and Richard Keating flew some atomic clocks around Earth. Some of the clocks went eastward and some went westward. From the reference frame of the center of Earth, the clock moving with Earth's rotation would have a greater speed than the plane moving westward. These clocks were then compared to an atomic clock at the US Naval Observatory, which was used as the reference clock. The theory predicted (using both general and special relativity) that the clock moving eastward would lose about 40 nanoseconds compared to the reference whereas the clock moving westward would gain about 275 nanoseconds. The clocks were measured to have a loss of 59 nanoseconds and a gain of 273 nanoseconds, respectively.

In March 2017 NASA started releasing data associated with a simpler version of the twin paradox. Astronauts Mark and Scott Kelly are identical twins. Although they have both been to space multiple times, Scott recently spent nearly a year in space. Because of this time in low-Earth orbit, zooming around Earth once every ninety minutes or so, Scott went from being six minutes younger than Mark to being six minutes and five milliseconds younger. In the grand scheme of things, however, this is not going to greatly affect Scott's life expectancy relative to his brother.

Perhaps the most useful (and most frequently cited) application of relativity is GPS. In order for a computer (such as your phone) to identify your location on Earth, it has to recognize where it is relative to some reference points. To do this in three dimensions, the device must communicate with at least four satellites. GPS satellites orbit Earth and are spaced out such that no matter where you are in the world, there should be at least four satellites visible to your device. Roughly speaking, the satellites say something like, "It is precisely 12:34:56.789012345 and I am moving this fast and at this location." Your phone receives this information a little bit after it was sent, so it's able to determine how far away the satellite is from the phone. Once it has enough references (from other satellites), the phone can pinpoint its precise location. Neither satellites nor the phone will work without taking into account the general relativistic effect of the satellites being farther from the center of Earth than the phone.

INVISIBLE HAND BRIDGE EVACUATION

WHEN Episode III, Chancellor Palpatine rescue mission	
WHERE Bridge of the *Invisible Hand*	
CHARACTERS Anakin Skywalker, Obi-Wan Kenobi, General Grievous	
PHYSICS CONCEPTS Pressure, temperature	

SHORT INTRODUCTION/BACKGROUND

Space is hostile to life. With no air, the ambient pressure in space is nearly zero, and the temperature is close to absolute zero. In order to survive there, ships need to maintain air pressure at levels close to atmospheric pressure on the surface of a planet. Having a metal container with a high-pressure gas inside of it while surrounded by the vacuum of space is stable unless that metal container is punctured. If a spaceship has a hole blown in its side, will there be a suctioning whoosh of air like there is in cartoons? How dangerous is a catastrophic failure of a window on the bridge of a spaceship? Could a person survive in space?

BACKSTORY

While on a rescue mission to free Chancellor Palpatine, Obi-Wan and Anakin must board the *Invisible Hand*, General Grievous's ship. During their escape Grievous captures all of them and brings them to the bridge. In the ensuing fight Grievous punctures a window and is hurled into space before bringing himself back to the exterior of the ship via a grappling hook. Obi-Wan, Anakin, and Chancellor Palpatine must hold on to secured objects to prevent their being pushed out as well. Ultimately, the blast shield closes and prevents the rest of the air from being sucked out of the bridge.

THE PHYSICS OF STAR WARS

Before we do any calculations, we must establish a few assumptions. Since the characters appear to be breathing without assistance and also appear to be reasonably comfortable as far as temperature goes, let us assume that the gas in the bridge is equivalent to air on Earth at 1 atmosphere (a unit of pressure) and 72°F. Let us also assume that the fractured window is about 12' × 6', or about seventy-two square feet. The room appears to be an irregular hexagon not dissimilar to a home plate from baseball. Given our assumptions for the dimensions of the window, an estimate of about seventy-five cubic feet of gas in the bridge is not unreasonable.

When gas is escaping a room in space, the amount of air in the room falls off exponentially. If you were inside the room when the air reached a density of around one-third normal levels, you'd feel as if you were standing atop Mount Everest. That is still survivable, but it will be difficult to breathe for an untrained person.

Given the numbers we assumed, it would take about a tenth of a second to reach this air density if the bridge room was sealed off from the rest of the ship. Since the doorway to the hall is open throughout, air from down the hall is able to rush in to replenish the air leaving the bridge. Overall, during the approximately fifteen seconds of the window being open, about two thousand cubic feet of air rushes out of the window. Given the size of the ship, that would most likely lead to a negligible change in ambient temperature and pressure.

But what about Grievous? Is he able to survive outside of the ship? If he were a purely mechanical being, that would be easy, but he has a biological brain, heart, and other vital organs. These must always be kept in a pressurized pouch because vital organs evolved to have counterpressure from surrounding organs and bones. Even if they were exposed to the vacuum of space, he probably would be okay. He is only outside for about ten seconds, which is brief enough for him to maintain consciousness and avoid death. Crucially, he gets back into the ship where pressures are closer to normal.

What about the temperature? Funnily enough, it is hard to lose energy through heat transfer in space. There are three mechanisms of heat transfer: convection, conduction, and radiation. Since space

EMPTY SPACE? NOT SO MUCH

Despite what you might think, space is not empty. According to quantum mechanics, particles are constantly being created and annihilated on microscopic scales. You may think, "Then why don't we see this?" Well, these particles only exist for a fraction of a fraction of a fraction of a second (a visible photon could exist for about 10–16 seconds). This is why it's safe to say space is empty. At that point, you might wonder why we care. It's because there are measurable effects caused by the microscopic particles, including decay of black holes via Hawking radiation, the Casimir force (attraction between neutral plates that are very close together), and Van der Waals forces. These all sound like science mumbo jumbo, but leading theories relate this to how geckos and spiders can climb sheer surfaces.

has no stuff in it, only radiative heat loss is possible. The rate of radiative heat loss increases with surface area and temperature. Assuming the biological brain Grievous has retained is of average size, it will drop in temperature one degree Fahrenheit every ninety seconds or so. This is not much of a temperature variation.

Water exposed to space experiences rapid evaporative cooling. Evaporation is just the process of molecules having enough energy to escape a liquid and float off as a gas. Typically, the pressure created by air molecules hitting the surface of a liquid prevents these molecules from escaping. At very low pressures, this is no longer the case, and evaporation happens significantly more easily. If Grievous's organs were completely exposed to the vacuum of space, he'd have to worry about them freezing, so the pressurized sac enclosing them is doubly important.

THE PHYSICS OF REAL LIFE

When you watch a scene like the one we've been discussing, it is hard not to imagine characters and objects on the bridge being pulled out into space by the breach in the ship's hull. In reality, the air in the room is pushing everything out.

Molecules move around quickly and randomly in gases such as air. When those molecules hit a wall, they bounce off the wall much like a ball. For an air molecule to escape through a hole, it has to run into the wall where the hole happens to be. Other molecules will continue to mosey around their room unaware that there is a hole until their path happens to bring them to it. This means that the image from a cartoon with lines of air being sucked out actually represents the velocity of the air molecules that happen to be heading for the hole.

Fortunately, catastrophes involving significant pressure changes are rare, though not unheard of. In 1966 Jim LeBlanc was in a vacuum chamber doing space suit testing for NASA. Due to a failure of the pressurization tube on his suit, he was exposed to a vacuum for 15–30 seconds. LeBlanc described feeling the saliva on his tongue boil before passing out. Due to the speed at which observers were able to return the room to normal pressure, he survived with no lasting damage.

On the other hand, in 1983 six workers were operating the Byford diving bell, a section of an oil rig off the coast of Scotland, which was at nine times atmospheric pressure when a catastrophic decompression event occurred. Due to a failure to follow protocols, the bell went from 9 atmospheres down to 1 atmosphere in a fraction of a second. Sadly, all but one of the workers died.

SPACE TRASH

"This bucket of bolts's never gonna get us past that blockade!"
—Princess Leia Organa (Episode V)

WHEN Episode V, escaping the Hoth asteroid field
WHERE Hoth asteroid field
CHARACTERS Han Solo, Chewbacca, C-3PO, Captain Needa
PHYSICS CONCEPTS Gravity

SHORT INTRODUCTION/BACKGROUND

Space is…very big. To get a better idea of just how big, consider a sphere centered at the Sun and stretching out to Neptune. Planet Earth and everything on it occupies 0.00000000000000028 percent of the volume of that sphere. Our entire solar system occupies only 0.0000000000000000000000009 percent of the space in the whole Milky Way. So, as we run out of room for certain things on Earth, it makes sense to look at the open areas above us for extra storage.

You might think that we could just throw all of our trash into space and not worry about it again. NASA and other organizations beg to differ and are already concerned about the amount of trash occupying our upper atmosphere. In a galaxy with far more species, planets, and space travel, would trash be a serious issue for the city, planet, and star system planners of Star Wars?

BACKSTORY

We can be pretty sure that the Star Wars galaxy hasn't developed a high-tech waste removal system because Luke and company find themselves in the Death Star's trash compactor in *A New Hope*. In *The Empire Strikes Back* we are introduced to the Galactic Empire's policy of jettisoning trash from large ships before heading to hyperspace. Is space big enough for that? We on Earth used to think the oceans were large enough to handle whatever we dumped into them, but we ended up with large islands of trash. Does the *Millennium Falcon* need to be just as concerned about space trash as it does asteroids or bounty hunters?

THE PHYSICS OF STAR WARS

When a Star Destroyer is preparing for the jump to hyperspace, it dumps its trash. If it were merely a question of space being big enough, this wouldn't be a problem. As discussed, space is really big, and the trash can't occupy all of it; it's just not possible. Trash is made of matter. Even if we treated literally all matter as trash, there is a great deal more empty space than matter. If trash is dumped in the same places over and over, however, you can run into a number of different issues.

In the event trash is dumped close to a planet, that trash will likely end up orbiting the planet. Whether you are on the surface of the planet or orbiting in a space ship, you would have to launch your trash away from the surface like a rocket to keep it from orbiting or falling back down. If populous planets such as Coruscant are putting their trash into space and not propelling it out of orbit, their skies will become so littered that it would be dangerous to fly. Even if there weren't enough trash to obstruct flying, objects left unattended in orbit will eventually fall to the surface. How long this takes depends on just how close to the planet an object is (with no resistance, it would orbit forever, but as you get closer to a planet, air resistance gets in the way).

When the trash did eventually fall to the surface, would that be a problem? Small pieces of trash (the size of a golf ball and smaller) would burn up in the atmosphere with little to no effect. Medium-sized pieces (from the size of a golf ball to the size of a bus) would burn up in spectacular fashion, looking like shooting stars. Large pieces of space trash (closer to the size of a London double-decker bus) would crash into the surface of the planet, destroying anything they collided with. In the case of extremely large pieces of space trash, not only would they destroy objects in and near their path, they could kick up enough dust and debris to change the climate on the planet for years, leading to mass extinctions. For comparison, the meteor that wiped out the dinosaurs was probably the size of Manhattan.

It sounds as if trash should probably just be spread out in space as much as possible. Maybe Kylo Ren could take a break from conquering the galaxy to establish a protocol about trash dumping within a certain radius of a planet, and Star Destroyers could let out a piece of trash here and there to keep it from coalescing. Even then, ships flying through that trail of trash could collide with it and damage themselves (or be destroyed completely if traveling fast enough). An array of space trash is just as dangerous to fly through as an asteroid field. Plus, if there is an area where people frequently jump to hyperspace, there might be a buildup of trash. Heavily traveled trade routes could slowly collect trash much like the litter on the side of highways. The most common trade routes between places such as Coruscant and Cato Neimoidia would need to have signs saying, "This section of space sponsored by the Padawans of Tatooine."

The difference between space trash and an asteroid field is that asteroid fields orbit stars. Without the stability of an orbit holding the pieces of space trash in their relative positions, gravity would ultimately pull the pieces together to form a mass. This might start as a small body, but would grow over time, perhaps becoming a landfill-style planet of sorts. That probably sounds gross, but at least then each sector could have its own trash "planet" to dispose of all its junk, and the existence of a dedicated location would hopefully keep ships from colliding with trash.

THE PHYSICS OF REAL LIFE

This idea of space trash collecting into a "planet" is not as far-fetched as one might think. We have seen a similar process on Earth with the creation of the Pacific trash vortex. Debris and trash dropped by ships drift on ocean currents that carry them into a region known as the North Pacific Gyre, where the currents form a vortex that traps the trash. Admittedly, there are no currents in space, but gravity can be a current surrogate, drawing the trash together.

Currently, we do not have a critical problem with space trash in our solar system. If you look up at the sky, it is not crowded out by space junk. But NASA and other space organizations are starting to be concerned about the stuff that's up there. At this time, there has only been one reported instance of a satellite colliding with space trash, but studies indicate that this number will increase exponentially with time. Already, the International Space Station (ISS) needs to dodge space trash on a regular basis.

Perhaps the most dramatic and famous instance of space trash was Skylab. A precursor to the ISS, Skylab was launched by NASA in 1973 and orbited the earth for about six years. When it was decommissioned, it fell back to Earth. It was intended to burn up in the atmosphere and for any larger bits to land in the ocean. Instead, many pieces crashed into a sparsely populated section of southwestern Australia. This ultimately resulted in an A$400 littering fine issued by the Australian government, which remained unpaid for decades; payment was ultimately crowdfunded by a radio DJ in San Francisco.

WHEN IS STAR WARS?

WHEN Episode IV, opening scene	
WHERE The galaxy	
CHARACTERS Before their time	
PHYSICS CONCEPTS Gravity, big bang theory	

SHORT INTRODUCTION/BACKGROUND

When watching a Star Wars film, you have to suspend your disbelief. Characters use magic brain powers to move things around, lightsabers are a thing, and there are civilizations spread over an entire galaxy. Despite all of these fantastical things, it's easy to think about this happening in our universe somewhere. This couldn't be our universe, though, could it?

BACKSTORY

From the beginning of *A New Hope*, we are told that Star Wars occurred "A long time ago in a galaxy far, far away..." Despite it being a long time ago, the galaxy displays signs of being very mature. It has billions of stars that have already formed. It has hundreds of intelligent species with interstellar transportation capabilities. But when is "a long time ago" in the history of the universe? Could the Star Wars universe be a parallel universe? Or maybe the whole saga predates the big bang altogether. Is that even possible?

THE PHYSICS OF STAR WARS

The best theories that we have indicate that our universe is about 13.7 billion years old. Assuming the Star Wars galaxy is in our universe, we need a few indicators to determine when in our universe's history Star Wars could occur.

The first galaxies were formed around a billion years after the big bang, so that cuts out a billion years. The films depict many star systems with mature planets and intelligent life. It took the solar system about 500 million years to form, and it formed 4.6 billion years ago, so it's reasonable to assume that Star Wars is about 5 billion years after the formation of the first galaxy. There are also fully formed multicellular creatures of many different shapes and kinds. It took about 2 billion years for single-celled organisms to evolve into multicellular organisms, and another billion years before those took the form of life that we could recognize as creatures.

Although it took billions of years for life to evolve on Earth, that does not mean the process would always take billions of years. The first eukaryotic cell is thought to have formed from a bacterium entering a prokaryotic cell and living symbiotically rather than being destroyed by its host cell. This was a random fluke that took millions of years to happen. If that event had happened on the first day of prokaryotic life, it could have shaved off a significant chunk of time for evolution.

Altogether, this means that Star Wars needs to be at least 9 billion years after the big bang. This leaves plenty of years before the current time (about 4.7 billion to be precise), so it could still count as "a long time ago," but it is certainly closer to now than to the big bang. Figuring out where the Star Wars galaxy is in the history of the universe is harder than trying to pinpoint planets in the history of their development. For instance, we see many terrestrial (rocky) planets and a few gas giants. A planet such as Mustafar is very volcanically active and has lava flowing all over its surface; Mustafar is probably in its earliest stages of development. In a similar fashion, Hoth could be an Earth-like planet in the heart of an ice age. It's more likely that it is just a cold planet far from its star, but Earth did undergo several ice ages.

Planet Earth was once a volcanic mess like Mustafar. Early terrestrial planets with a molten core go through a Mustafar-like phase. These planets need to dissipate heat, especially in early life. Volcanic eruptions aid in that goal and incidentally help create an atmosphere—molecules outgassed via these eruptions are trapped inside a collapsing sphere of dust (that is present from the collisions involved in planet formation). Eventually, the heat and dust clears out and the volcanism settles down.

So a long time ago is feasible, but if Star Wars does take place in our universe, why don't we find all the fictional elements such as collapsium, tibanna, and baradium? There are two possible explanations. Either they are not atomic in nature or they are larger, as yet undiscovered, elements. As of 2016, 118 elements have been discovered, and none of them are tibanna. If these elements exist, they have to be heavier than the heaviest discovered elements.

It is possible, though, that they are not atomic at all. We think of atoms as the building blocks of everything, but there are a whole host of particles that fall outside this definition. Tibanna could be a type of subatomic particle that is awaiting discovery.

THE PHYSICS OF REAL LIFE

The idea that tibanna and company are actually just heavy elements (i.e., elements with larger atomic numbers) that have yet to be discovered is unlikely. To explain why, we need to discuss radioactive decay.

Some elements have unstable arrangements of subatomic particles, which leads to loss of such particles in a process called radioactive decay (which results in less of the element in its original form existing). Unstable atoms with larger atomic numbers tend to decay more rapidly. Uranium (element 92 on the periodic table) is the largest element not discovered through laboratory synthesis and it has a half-life of sixty-nine to four and a half billion years (a half-life is how long it takes for half of a substance to decay; it can be a range of values depending on how the subatomic particles of the element are arranged). As you go up in number on the periodic table, atoms have significantly shorter half-lives. For instance, one element much larger than uranium has a half-life of about 0.7 milliseconds. So, tibanna would likely have a very small half-life and thus wouldn't stick around long enough for anyone to mine it (like they do on Cloud City).

SEABORGIUM AND OGANESSON

Glenn Seaborg had the honor of being the only living person to have an element named after him (number 106, seaborgium) until 2016 when oganesson was named after famed Russian-Armenian physicist Yuri Oganessian.

Still, the idea isn't completely impossible. In the 1960s Glenn Seaborg proposed an "island of stability" in larger nuclei, which could give new life to the undiscovered elements theory. Seaborg's idea relates to a concept in nuclear physics known as magic numbers. Just as electrons have shells that they can occupy while they orbit the nucleus, there are energy shells for the protons and neutrons inside the nucleus. Just as there are a certain number of electrons in a shell, which leads to a less reactive element (because the electrons have "filled" a shell), there are magic numbers of protons or neutrons that cause a nucleus to be extraordinarily stable. Although additional numbers have been proposed, the accepted magic numbers are 2, 8, 20, 28, 50, 82, and 126. This means that the not yet discovered element 126 could turn out to be very stable. It would be in the metal range of the periodic table, so it is unlikely to be tibanna gas, but who knows? Maybe it will be baradium.

It's also possible that Star Wars occurs in a parallel universe. Some interpretations of quantum mechanics indicate that our universe is just one of an infinity of universes. The idea of a multiverse is often explained by considering a block of Swiss cheese. Each universe is one of the holes in the cheese expanding outward. Unlike Swiss cheese, though, the multiverse itself is also expanding so none of the universes will ever collide with each other. In these other universes things could be identical (as far as the laws of physics are concerned) to our universe. Other theories indicate that perhaps in all of these other universes, fundamental constants such as π and the speed of light are actually different numbers. This could lead to wildly different behaviors than what we observe in our universe.

It is also possible that Star Wars took place in a time before the big bang. This may sound like an impossible suggestion, but the big bang just marks a point in history when our current laws of physics didn't work. It is possible a universe existed before the big bang; it would have been very hot and dense and collapsed upon itself. In this scenario, all of space and time were crumpled up together in an unrecognizable fashion. This is not science fiction—some theories suggest this. The biggest downside to these theories is that they will never be directly measurable unless we can travel to one of these universes or our universe collides with another universe.

PLANETARY SCIENCE

BINARY STARS

"You can't stop change any more than you can stop the suns from setting."
—Shmi (Episode I)

WHEN Episode IV, introduction to Luke Skywalker

WHERE Tatooine

CHARACTERS Luke Skywalker, Obi-Wan Kenobi

PHYSICS CONCEPTS Center of mass, gravity, Kepler's laws of orbits

SHORT INTRODUCTION/BACKGROUND

When we think of planets, we usually picture our solar system—one star with eight planets orbiting. Although a single star with many planets appears to be the most common configuration, there are more exotic systems. Many stars can orbit each other. There can be rogue planets, which don't seem to have a star. An example very relevant to Star Wars is a binary star system in which a planet orbits a pair of stars. You might think that it would be cool to have double the sunrises and sunsets, but it would actually be pretty hot living on a planet orbiting binary stars. Aside from disrupting planetary conditions, these systems also tend to be unstable and can't support life for very long.

BACKSTORY

In 1977 the opening scenes of the first Star Wars film awed audiences with an iconic shot of a Star Destroyer chasing a rebel blockade runner over the planet Tatooine. Tatooine went on to appear in nearly every Star Wars film; it is woven throughout the saga as it is the home of both Anakin and Luke Skywalker. Even so, the planet is introduced to us as the farthest point from any "bright center to the universe" and somewhere Luke can't wait to leave. Many of Tatooine's drawbacks as a home may be attributable to its twin suns, Tatoo I and Tatoo II (though they do provide a spectacular sunset).

THE PHYSICS OF STAR WARS

When we first see Tatooine's suns in Episode IV, they are both setting as our own sun does. The one closest to the horizon is a deep red; the other still has some yellow color. With that information alone we can assume the stars are about equal in mass, brightness, and type; we know this because stars are predictable. Larger stars tend to be hotter and more blue whereas smaller stars tend to be colder and more red. This means that if you can see the color of a star, you can also infer its temperature and mass within reasonable certainty. Our sun is a G-type star, which tend to be about 8,500–10,000°F and primarily emit visible light (the range of wavelengths of light that humans can see). Given that these stars look approximately the same size as our sun does to us, we can also estimate that Tatooine is about ninety-three million miles from its stars (roughly the same distance from Earth to its sun).

So, if Tatooine and its suns share many characteristics of Earth relative to its sun, why is Tatooine so different from Earth? For one thing, when you have two stars, you have double the sunlight. The novelty of having two suns will wear off quickly. Having two stars doubles the amount of radiation that the planet receives (at least when the stars are similar in type) as well as doubling the solar wind. Solar wind, when not deterred by a magnetic field, can blow away the atmosphere of a planet. Because Tatooine seems to have an intact atmosphere, we can conclude that there should be a molten iron (or other magnetic material) core inside of Tatooine generating a magnetic field strong enough to protect its atmosphere.

Another issue that could be affecting Tatooine is its orbit. Tatoo I and Tatoo II should orbit a point about halfway between them while Tatooine orbits around both of them. This is because any time you have two (or more) objects orbiting each other, they all orbit the collective center of mass of the system of objects. We think of our sun as sitting stationary in the center of our solar system because our sun is a thousand times more massive than the largest planet (Jupiter), so our solar system's center of mass is close to the sun. Because of the pull of Jupiter and the other planets, our sun is actually orbiting a point just barely outside its radius.

With that in mind, let's try to determine some information about Tatooine's orbit around its binary star system. In order to calculate the location of the center of mass of this solar system (called the barycenter), you have to calculate the weighted average of the locations of the individual masses. Since stars are significantly more massive than planets, we can get a good estimate of the location of the barycenter by finding the center of mass of the two stars.

We can use the mass of our sun to approximate the mass of each of the stars, but we also need to know how far apart these stars are on average. In order to find this distance, we can use Kepler's third law of planetary motion, which predicts orbital period (how long it takes for an object to complete a lap, so to speak) from the distance between the orbiting objects.

When we look at distant objects that are objectively the same size, their apparent size is directly proportional to how far away they are. Since every scene with the two stars depicts them as the same size, we can assume that they are much closer to each other than they are to Tatooine. If the distance to Tatooine from one of the stars was comparable to the distance between the stars, then the stars would appear to be radically different sizes. Since it would be hard to tell with a naked eye if the stars are about 1 percent different in size, the ratio of Tatooine's distance from the barycenter to the stars' distance from the barycenter is at most 100:1.

To recap, we know that Tatoo I and Tatoo II are approximately the same mass as our sun and each of them is approximately ninety-three million miles away from Tatooine (plus or minus about 930,000 miles). Using this information, we can predict both the orbital period of Tatooine around the stars as well as the period of the stars around each other. Tatooine will orbit the stars about once every eight and a half months and the stars will orbit each other about every six hours and eleven minutes. Sadly for Tatooine, stars orbiting each other that quickly typically are in a death spiral. More details on the death of binary stars follow in the Physics of Real Life section.

THE PHYSICS OF REAL LIFE

We know a fair amount about our solar system, but we are just learning about other planets around our galaxy. In 2009 NASA launched the Kepler mission in order to discover extrasolar planets (planets not in our solar system, sometimes shortened to exoplanets). This mission is using a space telescope (similar to the Hubble) to search for eclipses caused by a planet orbiting in front of a star. It does so by looking for dips in the brightness of stars—just as the sun is blocked out by the moon passing between the earth and the sun, the brightness of a distant star is decreased every time a planet moves between the viewer and the star. That means the telescope is only capable of detecting a limited number of planets (only those that pass between it and a star it is facing). Plus it's only scanning around 100,000 stars (of the approximately 100,000,000,000 stars in the Milky Way alone).

Nevertheless, in late 2009 the Kepler telescope discovered its first planet orbiting a binary star system. It was given the very dry name Kepler-16b, but the folks at the Smithsonian unofficially dubbed it "Tatooine" and the scientific community followed their lead. This is a bit of a misnomer because Kepler-16b is a gas giant orbiting stars of different types; the primary star is a K-type star whereas the secondary star is an M-type red dwarf. In other words, they are both smaller, less massive, and dimmer than our sun. By now we are aware of about twenty confirmed planets orbiting binary stars. Interestingly, many of them (including Kepler-16b) are near the habitable zone for their stars. This means that even if the planet itself is unable to sustain life (for example, because it's a gas giant), it might have a rocky moon or two that could.

Before you get too excited about the prospect of life on a Tatooine-like planet, let us explore the environment of a binary star system. Typically, a binary star system forms like most other stellar systems, via gravitational accretion (coming together) of space dust. For a binary system to form, the dust must accrete into two orbiting stars rather than a single primary star. Theories for how this happens have not been explicitly verified, but it seems likely that some kind of fragmentation happens to the dust cloud as it comes together.

Once the stars are formed, they will orbit each other for millions of years without much excitement. Eventually, though, they will move close enough together that one of the stars begins to leach material from the other. The star that is sucking the life out of the other is fittingly given the name vampire star. Once the vampire star has sucked enough material from its partner, the partner star will collapse. The mechanism of collapse depends on the type of the star, but one possibility is for the star to first go supernova (roughly speaking, for it to explode) and then collapse into a neutron star.

Living on a planet orbiting a star that goes supernova would not be a pleasant experience. Supernovae are the most cataclysmic events in the known universe. If our sun were to go supernova (don't worry, it's not the right type of star for that), it would heat Earth to a temperature hotter than that of our sun, and all life would be destroyed as Earth disintegrated. In fairness, all stars eventually die, and when our sun dies (in about 4.5 billion years), it will consume planet Earth with it.

CLIMATE OF PLANETS

WHEN All films

WHERE Hoth, Tatooine, Kamino, etc.

CHARACTERS All

PHYSICS CONCEPTS Orbits, gravity

SHORT INTRODUCTION/BACKGROUND

There are a number of factors that determine a planet's climate. For example, because of differing distances from the sun, Pluto is mostly a ball of nitrogen ice and rock while Mercury is a rocky planet with temperatures that can range hundreds of degrees from day to night. Some planets (such as Mercury) don't have an atmosphere, whereas gas giants (such as Jupiter) are mostly atmosphere. Without an atmosphere, wind, clouds, and weather are not possible. As far as we can tell, only Earth has a climate suitable for life on it, although some theories suggest that other planets in our solar system could have supported life when the Sun was younger. Looking outside our solar system, there is even more variability in planetary conditions. What other climate conditions are possible? Could we have found those conditions in our solar system at a different point in history?

BACKSTORY

The Star Wars movies display many different planets with varied climates, from the ice and snow of Hoth to the desert sands of Tatooine. There are gas giants, rocky planets, planets covered in magma, planets covered in water, planets with large caverns, forest moons, and so on. Would we expect to see such a range of climates in one galaxy? Are these climates even possible? Could they have supported life formation, or would life have had to migrate to them?

THE PHYSICS OF STAR WARS

Hoth has one of the harshest climates seen in the Star Wars films. In the opening crawl of *The Empire Strikes Back*, Hoth is described as an "ice planet." General Veers identifies Hoth as the sixth planet from its star. The farther away a planet is from its star, the less energy it will receive. This is because of the way energy spreads out as it moves away from a source.

The frigid temperatures on Hoth, given its distance from its star, are somewhat consistent with what we see in our solar system. Uranus and Neptune (the seventh and eighth planets from the Sun) are considered ice giants as they seem to be made completely of ice. This may sound like Hoth, but Hoth also has snow. Planets far from their sun typically consist of frozen nitrogen, methane, or other such molecules, not water. For there to be snow in our understanding of the word, Hoth would need to support liquid water. If Hoth were really as cold as most ice giants, the water would remain frozen on the surface and never enter the atmosphere to come back down as snow. It is possible that it is actually "snowing" nitrogen crystals, but even so the temperature on Hoth would be too cold to survive in a tauntaun.

PLANET OR NOT-A-PLANET?

Although Episode V identifies Hoth as a planet, could it be mislabeled? We know that Hoth has an asteroid field close by. Although the asteroids all seemed to be quite small compared to the planet, that is exactly what we previously thought about Pluto. We considered Pluto to be a planet because of its size compared to the many smaller, but similar ice balls in what is known as the Kuiper Belt. Eventually, scientists discovered Eris, an object in the Kuiper Belt larger than Pluto which led to the reclassification of Pluto as a dwarf planet.

HEAT CAPACITY

Anyone who has lived near an ocean or large lake knows that it can help maintain stable temperatures. Similarly, on an ocean planet, there would not be large fluctuations in temperature. This is due to water's ability to store thermal energy better than any other material. We call this property of a material its heat capacity.

Where Hoth is cold and wet, Tatooine is hot and dry. The most obvious reason for the difference is that it is receiving solar radiation from two stars. If this were the situation on Earth, southern Alaska's climate would become what the equator's climate is now, and the equator would become significantly hotter. It is hard to predict a precise temperature given the complicated feedback loops associated with the climate, but let's just say that moisture farming would likely be a very important occupation.

Finally, let's consider the planet Kamino. It is an ocean planet where (as far as we see) it is constantly raining and the surface is entirely water. It is unlikely that it always rains, but a planet that is completely oceanic would experience more powerful and more frequent storms than a terrestrial planet such as Earth. This is because the entire surface (all water) will receive solar radiation, leading to the formation of clouds. Those clouds will be moved around by winds created by hotter regions (currently receiving the solar radiation) and cooler regions. Clouds will reflect and block additional solar radiation, which will reduce cloud formation. Depending on its rotation speed, it's possible that Kamino could have a perpetual storm that moves around and is constantly fed by the ocean ahead of it, while it feeds the ocean behind it.

THE PHYSICS OF REAL LIFE

The climates of the planets in our solar system are quite varied. Mercury has long days and nights and almost no atmosphere, so while the side facing the sun heats up to around 800°F, the other side may be about −280°F (there is plenty of time to reach extreme temperatures but nothing to contain the energy—the heat is not trapped by an atmosphere). Venus, on the other hand, has more atmosphere than it knows what to do with. The clouds of Venus (which are made of sulfuric acid) reflect 75 percent of solar radiation; what does make it through the clouds is mostly trapped beneath them. This leads to temperatures on Venus that are actually higher (on average) than on Mercury.

Skipping Earth (hopefully we all have some familiarity with that planet's climate), Mars has reasonably predictable weather. Temperatures range between about a mild 70°F and a beyond-frigid −240°F; ice clouds can form around particles of dust from dust storms.

As for gas giants, because of their huge atmospheres and non-solid surfaces, scientists often refer to a threshold pressure (usually around the same pressure as Earth's atmosphere) to indicate where the atmosphere stops and the "surface" begins. Jupiter is famous for having a storm about the diameter of Earth known as the Great Red Spot; this storm has been raging at least 180 years.

THE POWER OF A STAR

A star is able to emit a certain amount of energy in a given amount of time. We call this the star's power. When a star radiates its energy, the energy moves away from the star more or less uniformly in all directions. To give a sense of the scale of this power, our sun gives off about 10^{25} kWh/day. The intensity of the energy decreases as it spreads out away from the star. When solar radiation reaches Earth, for example, it has an irradiance (the density of solar power) of about 1000 W/m². By the time it reaches Neptune, though, the irradiance is down to about 1.5 W/m². This means a solar panel on the earth could generate almost 700 times the power it could on Neptune.

Let's not restrict ourselves to the climates of the eight planets in our solar system. In examining exoplanets scientists have defined a region that is known as the habitable zone, informally called the Goldilocks zone. The habitable zone is far enough from a star to be not too hot, but not so far as to be too cold. There is variability in the brightness of (and thus heat delivered by) stars, so the habitable zone is not just around ninety-three million miles (the average distance from Earth to the Sun). As a star radiates energy, it goes equally in all directions forming a sphere of radiation. The amount of energy delivered to the surface of any planet decreases as you move farther from the star.

Another important aspect of the habitable zone is the idea that liquid water is fundamental to life on Earth. It could be that there are organisms that don't rely on liquid water, but we have no evidence for such life-forms at this time. Thus, the search for extraterrestrial life primarily focuses on planets that could have water.

At this time, counting Earth, there are only between ten and twenty-five known planets in the habitable zone of their stars. These planets could be completely inhospitable to life for other reasons. For instance, the closest of these planets, known as Proxima Centauri b, has an equilibrium temperature of about −38°F. Obviously, it is possible to survive these temperatures (as any resident of Buffalo, New York, can tell you), but it would not be an ideal candidate for life as we know it.

We know that life can develop from Earth-like conditions, but there could be life-forms that are so foreign to our concept of life that we are unable to conceptualize them. Already we know of organisms that are able to survive in environments hostile to most forms of life, including high acidity or alkalinity, temperatures above 130°F or below freezing, or areas high in UV radiation. We call these organisms extremophiles.

MOISTURE FARMING

WHEN Episode IV	
WHERE Tatooine	
CHARACTERS Luke Skywalker, Owen Lars, Beru Lars	
PHYSICS CONCEPTS Phase changes	

SHORT INTRODUCTION/BACKGROUND

As far as we know, water is necessary for life. When scientists consider the possibility of other life-forms in the universe, the presence of water (or lack thereof) on other planets is a main consideration. Still, if an environment does not provide easy access to water, existing life-forms can develop new ways of finding it—water can be pumped from the ground, moved in from other areas, extracted from the air, or reclaimed from other organisms, for example. Humans have been using several of these practices for much of our history, but clean water is still inaccessible to millions of people. How effective are various methods of getting water to dry areas?

BACKSTORY

Many of the planets in the Star Wars galaxy seem to represent different climates found on Earth. For the extremer climates (like Hoth or Mustafar), we don't necessarily see how organisms would adapt to the environment because there aren't many life forms. There is, however, one extreme environment that definitely sustains life: Tatooine. We know severe lack of water is a major problem for Tatooine because our first introduction to Luke Skywalker is on his uncle's "moisture farm." This farm offers the initial impetus for Luke to encounter C-3PO and R2-D2 and Luke's lack of attachment to the desolate planet makes him willing to leave home for an "idealistic crusade" with Obi-Wan. Luke might be bored with moisture farming, but aspects of his experience can help us better understand water issues that affect many people worldwide.

THE PHYSICS OF STAR WARS

Moisture farming involves extracting moisture from the air and condensing it into liquid. In order to do this, water vapor in the air needs to be converted from a gaseous state to a liquid state (a phase change). For a material to change phase, energy has to be added to it or removed from it. For example, for ice to melt into water, energy (typically heat) must be added to it. If you want to freeze water or condense water vapor, you must instead remove energy.

Uncle Owen's farm uses moisture vaporators to do this. The basic design of a vaporator requires ambient air flowing into the system. Inside, the air comes in contact with a coiled pipe that carries some kind of refrigerant (a liquid used to keep the pipe cold) which will cool down the air and cause some of the water to turn into liquid. The liquid is then collected and saved for later. In the condensation process the water will bring with it dust, ions, and other impurities from the air, so the water needs to be filtered.

All of these steps require power. Devices like this are primarily used in the desert, so they are generally powered by solar or wind energy. The vaporators shown in *A New Hope* have tall columns with a spinning piece on top, likely part of a wind power generation mechanism. Although we do not see solar panels in a form recognizable to us, they may use solar power as well. Tatooine is in a great location to take advantage of solar power from not one but two suns. Desert climates are also frequently windy because of

---

LIVING IN THE DESERT

Life in a very arid climate requires modifications in behavior and dress to stay healthy. For instance, the average human in normal conditions needs to drink around a gallon of water a day. In desert climates, this can increase to upward of three to four gallons. To help combat water loss, a person should cover exposed skin, nose, and mouth, and stay in the shade as much as possible.

large temperature variations and flat terrain that does little to block the wind.

THE PHYSICS OF REAL LIFE

The World Health Organization estimates that around a billion people don't have access to safe drinking water. Access to clean water would greatly reduce cases of cholera and other preventable diseases around the world. Lack of access almost always stems from financial issues rather than technological ones. Cheaper methods like wells can be unviable because groundwater is often contaminated by minerals from the soil or chemicals from nearby farms or factories. Transporting water may seem rudimentary, but it frequently requires expensive infrastructure. The moisture farming shown in Star Wars is actually very well designed for the types of communities that need clean water the most.

So, to what extent could we use moisture farming on Earth? How much water you can get out of the air depends directly on how much water is in the air. Relative humidity, a measure of how much water vapor is currently in the air, goes from 0 percent (indicating no water is in the air) to 100 percent (meaning the air is at its saturation point and no more water can be in the air without condensing to liquid on its own). You might expect that in a desert the humidity would be close to zero, but the average relative humidity in most deserts is around 25 percent.

A number of groups have developed technology to extract water from the air. Perhaps the cleverest involves a combination of atmospheric water generation and advertising: Peru's University of Engineering and Technology designed a billboard that pulls water from the air, filters it, and then delivers it to a cistern. The billboard was specifically made for Lima because of its proximity to both the Atacama Desert and the Pacific Ocean. Lima receives very little rainfall but has humidity upward of 90 percent on hot summer days. On a humid day the billboard is able to generate about twenty-five gallons of water.

A different design by a French company combines wind and solar power to accomplish a similar goal. Their wind turbines can generate 92 to 317 gallons of water per day depending on the humidity. Unfortunately, the wind turbines cost several hundreds of thousands of dollars, so poor rural communities can't afford them.

The National Peace Corps Association (a group for alumni of the Peace Corps) is working with a team of scientists at UC Berkeley and VICI-Labs to design a relatively inexpensive atmospheric water generation system. Their design (known as WaterSeer) should be able to produce around eleven gallons of water a day. This is significantly less than the French design, but the device costs only a few hundred dollars. The eleven-gallon projection is based on a prototype that is smaller than the final product. If expectations meet reality, one WaterSeer could provide safe drinking water for an entire family.

Many other organizations are trying to provide clean water to the world. Check out Evidence Action, the Uganda Water Project, and www.charitywater.org if you are interested in helping.

CLOUD CITY

"Lando conned somebody out of it."
—Han Solo (Episode V)

WHEN Episode V, immediately after the escape from Hoth	
WHERE Bespin, Cloud City	
CHARACTERS Lando Calrissian, Han Solo, Luke Skywalker, Princess Leia Organa, C-3PO, R2-D2, Chewbacca, Darth Vader, Lobot, Boba Fett	
PHYSICS CONCEPTS Air resistance, orbits	

SHORT INTRODUCTION/BACKGROUND

Cities of the future as depicted in science fiction are often floating in space (think the Jetsons). This may have to do with concerns that at some point life on Earth may not be viable. In the near future, we've got climate change to worry about; in the long-term there are all kinds of reasons we might need to relocate. Other planets tend to be less hospitable than Earth, so it makes sense that the imagination would go to an idealized, fully controllable environment like a floating home. Is that possible?

BACKSTORY

Star Wars provides us with a perfect example of a science fiction floating city: Cloud City. Cloud City has an atypical backstory. Floating above the surface of the planet Bespin, the city was specifically designed to harvest tibanna gas rather than to house a displaced population. Tibanna gas is used in all kinds of technology in the Star Wars galaxy, including, but not limited to, blasters and repulsorlifts (see the section on repulsorlifts for more). Being one of the few sources of the gas, Cloud City enjoys financial success from its mining operations. Does it make sense to harvest gas from a gas giant planet? What happens if you harvest the entire planet?

THE PHYSICS OF STAR WARS

It would be quite difficult and expensive to suspend a city above a planet. Probably the simplest way to accomplish this would be to have the city exist in a very low orbit around the planet. To maintain such an orbit, the city would have to move horizontally quickly enough that the planet curved away from it just as swiftly as the city moves around it. If the city can keep traveling at that speed it will maintain orbit and never crash down into the planet. When an object is in orbit, it's actually falling the entire time, it just moves to the side fast enough to always miss the surface of the planet.

Why is this the simplest solution? Let us consider the difficulties involved in instead hovering above the surface in one place. Assume for the sake of argument that Cloud City is an ellipsoid (a stretched-out sphere) that is about twenty kilometers across and five kilometers from top to bottom (this can give us an idea of the different forces required to maintain Cloud City's elevation). Also assume that the stratum of atmosphere Cloud City is in has about the same density of air as near Earth's surface. Finally, assume that the planet has a diameter of 73,322 miles (this is often cited as the diameter of Bespin).

With these assumptions, the city would need to orbit the planet at around 54,000 miles per hour. Satellites orbiting Earth typically travel around one-third of this speed, so that's fast, but not out-of-this-world fast. The drag force on something the size of Cloud City as it traveled this fast through air would be about 6.74×10^{18} pounds, or about one-seventh the force between Earth and the Moon.

Let's compare that to the force required to just hold up Cloud City. If we assume the entire city is made out of steel (to get a very rough estimate of the force required), then the necessary force would be 3.82×10^{25} pounds, or about ten thousand times the force between Earth and the Sun.

We've oversimplified the issue, but still, the force of drag is ten million times less than the lift force. This clearly indicates that it would be easier to have an orbiting city than to have a stationary/ hovering city. Of course, an orbiting city introduces its own issues, such as having to find where it is every time you visit.

Ignoring the difficulties involved in getting the city constructed and into orbit, wouldn't harvesting the gas eventually become a problem? After all, there is less and less of a planet to orbit as time goes by. The short answer is yes, you could mine the entire planet. As gas is mined, the mass and radius of Bespin would gradually decrease as to approximately maintain the planet's density. During this reduction in the planet's size, the pull of gravity would become weaker and weaker. As this happened, both the force to support the city and the force to keep it in orbit would decrease. Eventually, the mass of the city would be greater than the mass of the planet, and the planet would orbit the city which in turn would start orbiting the nearest star.

THE PHYSICS OF REAL LIFE

The idea of Cloud City probably sounds a little far-fetched, and rightfully so. As of the writing of this book, there are no floating cities around Earth, but there is an orbiting habitat designed for a few humans in the International Space Station (ISS). Just like the space station, if Cloud City were truly in orbit, everything in the city would appear weightless because everything would need to be in orbit in the same way. Unlike Cloud City though, the ISS does not have artificial gravity, so all of the people and objects on the ISS appear weightless. This is because everything in the ISS is in the same orbit as the station itself. Other aspects of Cloud City's setup are less fantastical.

The idea of creating a floating mine over a gas giant to harvest its gas has actually already been considered by NASA. In a report put out by NASA from April 2015, Bryan Palaszewski describes methods and vehicles that can be used to harvest gas from the outer planets of the solar system. He details the use of unmanned aerial vehicles that would harvest helium gas from the atmospheres of Neptune and Uranus before returning to a mother ship in orbit around the planets to deliver their payload. Perhaps the most interesting aspect of this report is the suggestion of the vehicles harvesting hydrogen alongside the helium with the goal of burning the hydrogen to power the rockets back up to the storage vehicle. The vehicles would actually be able to refuel themselves while they work.

Some exploration of the outer planets of the solar system has been done to determine whether it would be worth harvesting their gases. To this day we don't know much about the atmosphere and composition of the outer planets, but we have made some interesting measurements using the *Galileo* probe launched in 1989. While *Galileo* was orbiting Jupiter, it took a number of pictures and measurements of the surface and atmosphere of the planet.

The most dramatic measurement made by the *Galileo* satellite was the impact of Comet Shoemaker–Levy 9 with Jupiter in 1994. Fragments of this comet were about 1.24 miles in diameter and hit Jupiter at a speed of around 134,000 miles per hour. This collision is the most cataclysmic event ever measured in the solar system. The collision of Fragment G with the surface released in one instant the energy of about six hundred times the earth's nuclear arsenal. It left a crater on the surface of Jupiter about 7,456 miles across that is still visible to this day.

THE LOST PLANET

"Lost a planet Master Obi-Wan has. How embarrassing. How embarrassing."
—Yoda (Episode II)

WHEN Episode II, Obi-Wan searches for Kamino	
WHERE Coruscant, Jedi Temple	
CHARACTERS Yoda, Obi-Wan Kenobi, younglings	
PHYSICS CONCEPTS Gravitational lensing	

SHORT INTRODUCTION/BACKGROUND

Integral to the Star Wars galaxy as we see it in the films is the ability to easily and (somewhat) predictably travel from system to system. For example, on numerous occasions, the *Millennium Falcon*'s crew needs to quickly determine if they can make the jump to hyperspace from their current location. They seem to be able to locate planets, asteroid fields, etc. with ease. Humans have looked to the stars to aid in navigation for a long time, but mapping an entire galaxy with such detail is a completely different task. Do the movies give any indication as to how their navigation systems work? If so, does it have any basis in reality?

BACKSTORY

We get a brief glimpse into such technology in *Attack of the Clones* when Obi-Wan is chasing an assassin and is directed to a planet that does not appear in the Jedi archives. He claims the archives indicate gravity is exerting a force on the surrounding stars without any apparent object in the star map. Gravity is a very long-range force but is also an exceptionally weak force compared to other forces in the universe. If there is indeed a planet there, would its gravitational pull be noticeable on the surrounding stars? Can we detect the existence of a celestial body just by gravitational signatures?

THE PHYSICS OF STAR WARS

Yoda observes that the star and all the planets of the system in question have been erased from the archives, but what he refers to as "gravity's silhouette" remains. The idea of a silhouette created by gravity is a common occurrence in astrophysics. In the case of this system (which turns out to be the Kamino system), it would certainly exert a gravitational force on its stellar neighbors. This force would most likely not be significant enough to radically affect the orbit of the stars around the galaxy, but it would provide measurable perturbations on the orbit.

Depending on the proximity of any neighboring systems and the mass of the star at the heart of the Kamino system, the strength of the effects could vary quite a bit. Gravitational attraction is larger for massive objects, but decreases as objects move away from each other. This means that for a measurable effect to exist, the star at the center of the Kamino system would need to be massive or it would need to be located moderately close to its nearest neighbors. Our intuitive definitions of massive and close may need to be recalibrated for astronomical sizes, though.

If you consider our solar system as an example, the nearest star system is Alpha Centauri. This is actually a collection of stars featuring a binary star system gravitationally bound to a fainter red dwarf. The binary system is made up of two stars that are very similar to our sun and are about 4.37 light-years away. The gravitational force that either of these stars exerts on our sun is three times greater than the force Pluto exerts on the sun.

So, a gravitational silhouette is feasible, but how would one go about detecting it? If you are a Jedi, that could just come with your force powers. Jedi can exert a force on or sense changes to remote objects. Maybe when the Death Star blows up Alderaan, Obi-Wan is able to sense a change in the gravitational forces and sound waves near that planet and that's how he is aware of voices "crying out in terror." The maps in the Jedi archives could be the work of Jedi over the ages sensing changes in the force and recording them.

A more realistic element that could play a role in mapping the universe is light. In the section on binary stars, we discussed how telescopes like Kepler can detect differences in star brightness in order to locate planets. Einstein's theory of relativity suggests that gravity can redirect light, a phenomenon known as gravitational lensing. Just as the lenses in your glasses (or contacts) bend light so that you can see, so does gravity. Since massive bodies (such as black holes or clusters of galaxies) bend space itself (more on this in the following section), the shortest distance for light to travel could involve bending around a massive object.

We can see the effects of this by looking at a large cluster of galaxies such as Abell 2218, photos of which show a circular rim of distortions. These distorted images are actually light from the galaxies on the far side of Abell 2218 being bent around this galactic cluster. This is just one of the many ways we can use light to map the universe.

THE PHYSICS OF REAL LIFE

Scientists have used gravity's silhouette to make a number of discoveries in astronomy throughout history. The earliest of these was heliocentricity, or the idea that everything in the solar system revolves around the sun. Although Copernicus first detailed the theory in the sixteenth century, it was not generally accepted until Newton described the universal law of gravity a century later. With Newton's law of gravity, we are generally able to make accurate predictions about where the planets will be in the sky from year to year. Interestingly enough, Mercury always seemed to be a little out of place based on Newton's predictions, and it was not for several centuries that we were able to figure out why.

In 1915 Einstein published his theory of general relativity suggesting that space-time is curved by massive bodies. That is to say, space itself bends around any object with mass. This bending of space is what gives rise to gravitational forces. The classic analogy involves stretching out a sheet and placing a bowling ball and a golf ball on it. Because the sheet is pulled down around the bowling ball, the golf ball is always attracted to it.

In 1915 Einstein's suggestion was radical, and most scientists were skeptical. One of the primary factors that convinced the scientific community that Einstein's theory is correct was that it perfectly explained the small variation in Mercury's orbit. Using the idea that space-time is curved due to the large mass of the Sun, time on Mercury would flow just slightly slower, which would explain why, from the perspective of people on Earth, Mercury appeared to be in the wrong place.

A related aspect of this is dark matter. Dark matter makes up about 27 percent of the mass of the known universe, but it does not interact with light, making it invisible to all light-sensitive telescopes. It does interact gravitationally, though. The reason we suspect dark matter's existence is due to its gravitational signature left behind in cosmological interactions.

Perhaps the most famous example of this is the Bullet Cluster. When you look at the Bullet Cluster, you can see what appears to be the results of a collision of two clusters of galaxies. If you measure where the center of mass is for the normal matter, you will find that it cannot explain the observable gravitational lensing. This discrepancy is explained away by suggesting that there must be a large amount of additional mass that cannot be seen—hence, dark matter.

Another cosmological object only detected via gravitational effects is the black hole. First explained by Karl Schwarzschild through general relativity, the theory of black holes has now been accepted as correct. When we measure the locations of stars orbiting the center of the Milky Way, they all appear to orbit the same spot in space. The problem is, like Obi-Wan, we don't see an object there. Using the laws governing orbits, the mass being orbited should be millions of times greater than that of our sun. This points to the existence of a black hole at the center of the galaxy.

PLANET-BASED TRANSPORTATION

REPULSORLIFTS

"Wait, there's something dead ahead on the scanner. It looks like our droid...hit the accelerator."
—Luke Skywalker (Episode IV)

WHEN Episode IV, search for C-3PO and R2-D2	
WHERE Tatooine, Owen and Beru Lars's moisture farm	
CHARACTERS Luke Skywalker	
PHYSICS CONCEPTS Newton's laws, gravity, electromagnetism, torque	

SHORT INTRODUCTION/BACKGROUND

In the Star Wars galaxy antigravity technology appears everywhere. Jabba's barge, Han Solo's carbonite-frozen body, and Luke's landspeeder are just a few examples of objects that we see levitating. It is easy to imagine how this technology works conceptually. The force of gravity holds all of us on the ground, so if we could get rid of gravity's effects on particular objects, they'd levitate. Is this possible? How could the height of the levitation be controlled? A lot of research has been done on this topic, but where's my hover car?

BACKSTORY

The first hover vehicle that we encounter in Star Wars is Luke's landspeeder. Watching that vehicle glide smoothly over the Tatooine desert shows how powerful antigravity technology could be; driving over the rough terrain would be very bumpy and much slower in a traditional wheeled vehicle. Eventually this technology is given a name: repulsorlift. A high-speed example of repulsorlift capabilities is shown on the forest moon of Endor in Episode VI and in the prequels we see this technology used by the Galactic Senate to float representatives from different locations in a large chamber.

THE PHYSICS OF STAR WARS

How could the repulsorlift eliminate the force of gravity? Of the four fundamental forces in physics, physicists least understand gravity and how it fits into the universe. The other three forces conform to a unified general theory, but including gravity in this theory has proven to be very difficult. Our understanding of how to eliminate gravity is rudimentary at best so could we eliminate the mass of the object instead? Alas, no. This would also require removing all the object's energy and thus eliminating the object itself (which would violate the rule of conservation of mass). The less simple-sounding option is to reverse the warping of space-time around nearby objects.

Removing the bending of space-time around an object is, surprisingly, not out of the question. Every object with mass warps space-time; space itself is distorted close to the object. To put this in less abstract terms, picture a big hill with a ball on the side of it. When you let go of the ball, it will roll down. If you were to dig out a small ledge so that the ground directly beneath the ball were flat, the ball would stay in place even as other balls on the hill continued to roll down. If we could locally bend space-time, we could cause something to hover in place even as everything else is affected by gravity. Larger objects naturally do this, hence why we feel gravitationally attracted to the surface of Earth but not to the book in our hands. If we could find a way to smooth out the bending in space-time in a localized region, then an object could ignore the force of gravity.

There is one other option. You could create a force that balances gravity rather than eliminating it. This sounds simple, but would require some very precise feedback loops. For instance, consider Jabba's sail barge. With no passengers or cargo on it, it's easy to balance gravity's force on it, but as Jabba boards and moves from one side to the other, the repulsorlifts on the underside would have to adjust how much force they apply. When Jabba moves back and forth, different torques are exerted on the ship; without constant adjustments of the balancing forces, the barge would capsize. If that is an unintuitive scenario, imagine trying to get in a canoe without a counterbalancing force to prevent it from turning over.

Even if we overcome the issue with the feedback loops, we haven't established how to produce the upward counterbalancing forces. Current technology requires us to expel gas at exceptionally large speeds to propel rockets upward. In principle, rockets could create just enough thrust to balance the force of gravity, allowing our ships to float in space. This would, however, require tremendous amounts of fuel, and it would be impossible to stand anywhere close to a vehicle being supported this way as the expelled gas is very hot (imagine standing under a space shuttle as it launches).

THE PHYSICS OF REAL LIFE

Since we've established how impractical or potentially impossible these technologies are, you might think there's a bad outlook for repulsorlift-like inventions. In fact, we already have transports that can levitate above the ground. Maglev trains follow a track, but instead of steel wheels on steel beams, they are supported and propelled by electromagnets. If you remember playing with magnets as a kid, you know that one side of a magnet repels a second magnet, while the other side attracts the second magnet. We can use magnet repulsion to hold a train up in the air and slightly modify the strength of the field to drive the vehicle forward.

Many designs could accomplish this, but let's focus on a specific example: electrodynamic suspension used by the Central Japan Railway Company. Consider the following diagram (where you are viewing the train as it approaches you). The train is currently in a position where it will feel an upward force from the magnets, keeping the train supported. Thrust can be applied by changing the magnets surrounding the train. With the train moving to the right (as shown on the right side of the following diagram), it will be more attracted to the magnets of opposite polarization directly in front of it and repelled by the magnets adjacent to it. As soon as it has advanced to the next position in this diagram, the magnets of the track reverse polarization, thus attracting the train farther forward and repelling it from its current position.

This type of technology has also been advanced by billionaire Elon Musk with his proposal for Hyperloop which is essentially a maglev train in a tube. The tube part of the design is intended to boost efficiency; a vacuum can be created in the tube. The Hyperloop trains would thus be free of contact friction forces (which maglev trains avoid) as well as air resistance forces. This would provide a nearly force-free ride; once the train is moving, it would require no thrust to maintain its velocity. This idea is still in its infancy, but it has the potential to travel between Stockholm and Helsinki (a distance of about 250 miles as the crow flies) in half an hour, or about half the time of an airplane flight.

There is a third demonstrated way to stably levitate an object: by using superconductors. When a material is a superconductor, it is able to expel all magnetic fields and enter a state we call flux pinning. This means that the superconductor will do whatever it can to maintain its orientation relative to a nearby magnetic field. If you put a superconductor above a magnet, it will remain exactly where you put it regardless of its orientation. If you lift it up a little bit, it will sit there. If you lower it a bit, it will stay there. If you tilt it in place, it remains tilted. What's the catch, you ask? Well, as of now, superconductors can only exist at very low temperatures. Current "high temperature" superconductors can get up to about −211°F before losing their superconductivity. Still, the temperature threshold has increased over the years. In the 1970s we were excited by superconductors at −400°F. The announcement of the

−220°F superconductor was first published in 1993; later versions have had temperatures as high as −94°F, but those versions are yet to be experimentally verified.

PODRACERS

"When the storm is over, you can see my racer. I'm building a Podracer!"
—Anakin Skywalker (Episode I)

WHEN Episode I, Boonta Eve Classic	
WHERE Tatooine	
CHARACTERS Anakin Skywalker, Sebulba	
PHYSICS CONCEPTS Tension	

SHORT INTRODUCTION/BACKGROUND
It appears that a long time ago in a galaxy far, far away, racing was just as popular a sport as it is presently in our world. In Episode I we are introduced to the exciting twists and turns of podracing. Every year, the best podracers on Tatooine converge for the Boonta Eve Classic. Podracing appears to be a mix between F1 racing and dogfighting. Could we build a podracer with our current technology?

BACKSTORY
While traveling between Naboo and the Galactic Senate, Queen Amidala hides out on Tatooine while the Jedi look for a replacement part for their ship. On Tatooine, Qui-Gon and Obi-Wan encounter Anakin Skywalker, a child who has a midi-chlorian count higher than that of Yoda. In order to win Anakin's freedom from slavery, Qui-Gon makes a bet with Anakin's master, Watto, that the boy will win the Boonta Eve Classic. Anakin has built his

podracer from the ground up in his spare time whereas most of the other racers have purchased their ships from companies. Ultimately, Anakin wins his freedom and proves himself to be one of the best pilots Obi-Wan has ever known.

THE PHYSICS OF STAR WARS

The apparent standard design for a podracer is that of a cockpit attached to two giant engines via long cables. The engines sit in front and come in a number of different shapes and sizes while the cockpit is located in the back. Most podracers at the Boonta Eve Classic have two cylindrical engines that look similar to traditional jet engines on aircraft here on Earth. A few notable exceptions exist, though. For instance, Ben Quadinaros has a four-engine model and Neva Kee has a single piece (no cables) with the cockpit in front. Do these designs even make sense?

When walking near an airplane, airport employees worry about two primary danger zones: the inlet danger zone (the front of the engine; dangerous because one could be sucked in) and the exhaust danger zone (the back of the engine; dangerous because air is being blown out). For the purposes of a podracer, there is less concern about the inlet danger zone because podracers have fins in front of the engines used for steering, which double as protectors. The real worry for podracers is the exhaust zone of their engines. A typical Boeing engine creates the equivalent of a category three hurricane (120-mile-per-hour winds) while idling; as well, the exhausted air is upward of 100°F. At full thrust, this engine easily exceeds a category five hurricane (155-mile-per-hour winds) even at distances of two hundred feet. Judging by the size of Anakin's cockpit, the most generous estimate for the distance between his engines and the cockpit is fifty feet. If the podracers' engines produce similar exhausts, he's well within the category five range and constantly being hit with 100°F winds that could knock over a house.

Of course, the podracers' engines aren't necessarily comparable to jet engines. Maybe they have a special engine design that sidesteps these issues. Is that feasible? In order to create a forward force on the podracer, a force must be exerted backward on something external to the vehicle. A jet engine expels exhaust backward to create forward thrust. It is possible that these engines are exerting a reverse force on something else. This does not seem especially reasonable since they have spinning blades, which would indeed exert a rearward force on the air; plus, the engines are explicitly shown with flaming exhaust coming out their backs. You can even see the distortions in the air caused by the heat from the engines as the camera pans around the podracers before the race begins.

There is at least one other design element that warrants a discussion. The podracers have energy binders that create purple lightning between the engines. The energy binders hold the engines in the appropriate locations. If this purple electrical energy were caused by the ionization of the air between the engines, it would require quite a bit of voltage. Air requires about 3,000,000 volts per meter to ionize, so there might be about 5 million volts between the two engines. Assuming Jar Jar's body is as conductive as a human body, this voltage would stop his heart and burn his skin, not just numb his tongue.

THE PHYSICS OF REAL LIFE

Is it possible to construct a podracer in real life? The short answer is maybe. The engines would need to have some sort of support to elevate them prior to takeoff, but they could remain off the ground once running if they had a flat, wing-like surface attached to them to create lift. As you may have noticed, this is roughly describing a jet airplane. In some sense, a podracer is just a uniquely designed airplane that flies very close to the ground.

Much like podracing, Formula 1 (and other types of vehicle racing) has evolved over the years so that most teams have cars with similar designs. Innovators who are willing to take a risk with design (like Neva Kee and Ben Quadinaros in *The Phantom*

Menace) push the field forward. Sometimes the designs are a bust, but sometimes they revolutionize the field. One of the most influential pioneers in F1 racing was Colin Chapman. In 1962 his design of the Lotus 25 introduced a chassis that replaced the previously used tubular frame. This allowed the body of the car to be structurally and aerodynamically more efficient, also cutting down its weight. In 1970 Chapman further revolutionized the design of F1 vehicles by placing the air intake above the driver's head and making the radiators side pods (rather than in the front nose of the vehicle). With these design changes, Chapman's car moved fourteen kilometers per hour (8.5 mph) faster on straightaways than other cars of the time. Soon every car followed suit.

TRAVELING THROUGH THE PLANET'S CORE

"Wesa give yousa una bongo. Da speedest way tooda Naboo tis goen through da core. Now go."
—Boss Nass (Episode I)

WHEN Episode I, traveling between Otoh Gunga and Theed

WHERE Naboo Abyss

CHARACTERS Qui-Gon Jinn, Obi-Wan Kenobi, Jar Jar Binks

PHYSICS CONCEPTS Gravity, pressure

SHORT INTRODUCTION/BACKGROUND
Just about every child has wondered, "Can I dig a hole so deep I'll go all the way through?" It doesn't seem entirely out of the question (assuming you found a way to avoid the molten core). Would it be possible on a gas giant, though? Do all terrestrial planets have molten cores? If we could cut through a planet's core, would that be the fastest way to the other side?

BACKSTORY

In *The Phantom Menace*, Qui-Gon Jinn and Obi-Wan Kenobi are dispatched to investigate the Trade Federation's invasion of the planet Naboo and end up stranded on the planet with no transportation. To the chagrin of many fans, they are guided around by one Jar Jar Binks. With assistance from Jar Jar and other Gungans (Jar Jar's species), the Jedi quickly travel to the other side of the planet in a bongo (a kind of submarine). The trip is so fast because instead of a molten core, Naboo is filled with mostly water. Does this make sense?

THE PHYSICS OF STAR WARS

The core of Naboo is shown as a series of caverns filled with water. This water is clearly connected to the oceans on the surface of either side of the planet given that Qui-Gon, Obi-Wan, and Jar Jar are able to travel from one ocean to the other, but if so, this would violate known scientific principles.

There is a limit to how deep a liquid ocean can be. Whether we have liquid, solid, or gaseous water molecules depends on pressure and temperature. Bodies of water have layers, and each layer has a different temperature. The top layer is the hottest since it receives direct sunlight. There is a layer below the top where currents mix the hotter water above with the cooler water below. Below the currents, the water gets progressively colder as it gets deeper. The deepest parts of the oceans on Earth, for instance, are usually around 32–41°F.

As the depth increases in a body of water, so does the pressure. Perhaps you have heard of divers who come up too quickly and get a condition called the bends. This is caused by moving through a change in water pressure too quickly for the body to stabilize; bubbles can form in the bloodstream and joints, causing anything from mild pain to death.

The increase in water pressure occurs because there is more water above weighing on the deeper water. This is not the full story, though. In a planet that has water going from its surface to its core and then back up to the opposite surface, the water pressure would have slightly more complicated effects. Roughly speaking, the deeper you go in the direction of the center of a planet, the pull of gravity gets weaker. At the perfect center of the planet, the force of gravity reaches zero. This is because the force of gravity only cares about material between you and the core. The consequence of this is the pressure increases less as you near the center of the planet, but reaches a maximum at the core.

The phase diagram of water is a chart that shows at what temperature and pressure water will freeze or evaporate. If you scan this chart in the areas around 32°F, you will see that the pressure at which water transforms to ice is around 92,000 psi, or about six thousand times that of standard atmospheric pressure on Earth. This pressure corresponds to a planet with radius close to 1,200 miles. This is half the radius of Earth and way deeper than the deepest point in its oceans (the Mariana Trench, the deepest location in the ocean, is about 6.8 miles deep). This sets a limit to how large Naboo can be before the water in its core changes into ice.

Some sources indicate that Naboo has a radius of about 7,456 miles. If this were the case, water at the core of the planet would freeze under the pressure from the water above it. If this radius is correct, there are a few possibilities that can still explain what we see in the film. Perhaps when the Gungans say, "Through the core," they don't literally mean through the center of the planet. It could still be faster to go deep into the ocean and around a central core of ice or rock. It could also be that the oceans are not made of water but are another material that mimics the appearance of water, but would be a liquid at these depths.

Another explanation would be heat generated inside the planet. If the oceans on Naboo are 7,456 miles deep, heat could keep the water in its liquid phase. Although we have not done experiments to fully verify exactly how hot the water would need to be in order to stay a liquid at such high pressures, the best estimates that we have indicate the water would need to be at least 1,340°F. This is approximately one-sixth the surface temperature of our sun, so it would be surprising if the bongo had sufficient shielding from the heat.

These numbers also change depending on the salinity of water, but not enough to keep the water in a liquid state. None of these explanations seem satisfactory; it is more reasonable that the radius of Naboo is at most about 2,000 miles as this would allow for liquid water all the way through its core.

THE PHYSICS OF REAL LIFE

Even though we don't have oceans going all the way through the core, many people have asked what would happen if we drilled a hole through the planet. Could we make it out the other side?

There are a few things standing in our way of trying to make an express elevator shaft to the opposite side of Earth. First of all, the center of Earth is a big ball of molten iron at a temperature of about 10,000°F, or about the temperature of the surface of our sun. This core is assumed to be present in all rocky planets, but direct measurements haven't been done on Earth, much less on other planets. Similarly, direct measurements of the cores of gas giants have not been done, but we theorize that their cores are iron-nickel rock surrounded by an exotic fluid known as liquid metallic hydrogen.

Passing through a pool of liquid iron does not sound like a pleasant experience to me. You could survive a very brief dip in such a hot pool of molten metal (as demonstrated by *MythBusters*) if you first soaked yourself in water. The water would evaporate and create a thin, protective barrier between you and the liquid metal (called the Leidenfrost effect), but this barrier only lasts for a fraction of a second, so don't rely on this to keep yourself safe all the way through the core.

Let us pretend for a second that the core was not super-hot and made of liquid metal and we could drill a hole through it to the other side. We would still encounter problems dropping down into the shaft.

Worst-case scenario, no adjustments are made to the shaft and you just try to fall through. Ignoring air resistance, when you reached the middle of the earth you would be going about 22,000 miles per hour, or about Mach 29 (twenty-nine times the speed of sound). If we factor in air resistance, your speed would cap off at 122 miles per hour. This is a reasonable speed, but if you were to drift to the side and bump into the wall of the shaft, it would end poorly.

Speaking of encountering the side of the shaft, we need to make sure that you are falling in a perfectly vertical line. Even a slight sideways velocity would cause you to hit the wall eventually since you have such a long way to fall. You may imagine you can just jump straight down, but that ignores how quickly you would be moving due to the rotation of Earth. All points of Earth perform one full rotation each day. This is true for both the surface and the core. Since the core has less distance to travel than the surface, there is quite a difference in speed in those locations. This means that when you jump into the hole, you are actually traveling east at a speed of 1,038 miles per hour relative to the core and you'd hit the eastern wall of the shaft before going very far.

Scientists have tested this in a few very deep mineshafts. For instance, the Tamarack mine shaft near Lake Superior is about a mile deep. Miners lowered plumb lines (metal spheres attached to a string, in this case) from the top of the shaft down to the bottom to determine the vertical distance. None of the plumb lines they dropped ever made it to the bottom. They never found many of them, but the ones that they did find were embedded in the timbers supporting the shaft on its eastern wall.

Even if you managed to overcome all of these obstacles (say by starting at the North Pole and jumping to the South Pole), would this even be faster? It turns out that if you run the numbers, an object orbiting a planet will reach the other side in exactly the same amount of time that an object being dropped through the center will take. This seems impossible, but despite the significantly shorter distance, the object falling through the planet will experience less and less gravity as it approaches the center of the planet. On its way out the other side, it will also be slowed down until it reaches the other side and comes to rest.

HOVER CAR TRAFFIC

"That was wacky! I almost lost you in the traffic."
—Anakin Skywalker (Episode II)

WHEN Episode II, pursuit of Zam

WHERE Coruscant

CHARACTERS Obi-Wan Kenobi, Anakin Skywalker, Zam Wesell

PHYSICS CONCEPTS Speed, acceleration

SHORT INTRODUCTION/BACKGROUND

Everybody hates traffic. It is a drain on time, mental energy, gasoline, and the environment. There was a time when it seemed that hover cars might revolutionize the traffic issue, but such inventions have mostly taken a backseat to things like self-driving cars and Lyft. With hundreds of thousands of cars unnecessarily producing extra carbon emissions every day (in general, but especially via traffic), nearly everyone has a vested interest in reducing traffic. So, what are the causes of traffic? What technological advances can we make to avoid traffic? Would hover cars improve the situation at all?

BACKSTORY

Near the opening of *Attack of the Clones*, Obi-Wan and Anakin pursue a suspected assassin through the skies of Coruscant.

Throughout the city, there appear to be "roadways," not in the traditional sense of a paved surface, but in layers of lines of hover cars. It seems that even with the many layers of cars in the city, there is still congestion. Later in Episode II, we see hover shuttles, taxis, and miscellaneous other standard metropolitan vehicles, just all of the hover variety. Coruscant appears to have all the hallmarks of traffic problems, they just occur in the air.

THE PHYSICS OF STAR WARS

In order to minimize traffic, there needs to be organization amongst the drivers on the "roads." In Coruscant, it appears that there are at least ten layers of traffic that Obi-Wan and Anakin fly through in their pursuit of the assassin. It could make sense to have the bottom row be akin to the "exit lane" and the higher lanes the expressways. The issue with this idea is that there appear to be landing zones at many different elevations throughout the city. When Anakin grabs a hover car at the start of the chase, he is already more than fifty stories up. There might be a regulation where turnoffs can only happen at certain elevations such that there would be a lower expressway, a middle expressway, and a higher expressway. There is no indication that this is the case in the movies, but it is a possible suggestion for a way to help control the issues with traffic.

There does seem to be some form of regulation keeping the cars in a rough cylindrical shape as they go parallel to the ground. While Obi-Wan is holding on to Zam's probe droid, he needs to dodge oncoming traffic which comes at him from many different elevations. If we were to apply similar traffic models from our galaxy upon the cylindrical hover car lane, one would expect the center of the tube to be the "fast lane" of traffic while the boundary layers are the slower people. This may sound a bit strange at first, but if you think about the interstate, the lanes that (typically) merge with exit lanes are on the right whereas the fast lanes are on the left. The fast lanes try to have as few merges as possible to minimize slowdowns. If you extend this to a cylinder, hover cars would potentially merge from the right, left, top, bottom, and everywhere in between. Thus, the "lane" farthest from the merges would be right down the center of the cylinder.

Switching between one cylinder of traffic to another cylinder of traffic would not be overly difficult since you pretty much have all the space you need to do the merge. That is, unless you opt to change lanes in the way that Obi-Wan or Anakin did. Both of them experienced a fall from one layer of traffic to the other. In Obi-Wan's case, Anakin met him in the hover car before resuming the pursuit. Anakin jumped out of their hover car to board Zam's hover car. If this were to happen in our galaxy, Anakin would fare much worse than Obi-Wan.

In the case where Obi-Wan is falling, it appears that he eventually reaches terminal velocity. In the script, it indicates he fell fifty stories, but that could just be a number to indicate "a lot of" stories. If we say he fell fifty stories and used the standard that commercial buildings tend to have twelve feet per story, Obi-Wan fell about six hundred feet (or about half of the Eiffel Tower). After this fall, he'll be traveling about 200 feet per second (assuming no air resistance), but this is higher than a human's terminal velocity (with flat belly), so he would probably be traveling closer to terminal velocity of 121 miles per hour. Falling this fast without a parachute sounds like it cannot end well. Anakin saves the day, though, by catching him. This "catch" shows Anakin matching Obi-Wan's downward velocity so that he lightly lands on the hover car before accelerating his way back to chase speeds.

Anakin on the other hand does not have a driver trying to catch him. If Anakin were going at 121 miles per hour and hit a hover car, he would not fare so well. Given the amount the hover car reacts to his hitting it, it would be much better than hitting the ground for sure. Hitting the hover car, if we assume that he is stopped in about half a second (given the hover car's dip, this wouldn't be too unreasonable), Anakin would feel a force approximately the same as an alligator bite as a blunt force to his body. This probably is survivable, but he'd certainly have a broken rib or two. The script technically only calls for him to fall five stories (admittedly the movie indicates otherwise), but if he were to only fall five stories, he'd only be going forty-two miles per hour and would feel like as if he'd been given a tough punch when he landed; certainly he'd survive with only minor bruises.

THE PHYSICS OF REAL LIFE

In our galaxy, we don't have hover cars, but we do have plenty of traffic. The closest thing we have to hover cars and the traffic associated with that would be airplanes and airports. As of right now, we don't have enough airplanes flying across the sky to need to worry so much about traffic; most airplanes fly along specified paths between two destinations. With only minor variations, the path a plane takes from New York to L.A. will mostly follow the same route no matter the airline or time of day. Obviously, there are times when a plane will change course to avoid a storm or other weather event. You can see this in action if you look at flight tracker maps. Since flight data is public, there are websites that show the current location of all registered airplanes in the sky at any given time. Since these planes are all following the same flight path, you could argue that we have traffic lanes for flying vehicles. Those lanes are not overcrowded until you near a merge point (an airport), where the planes all have to line up and wait their turn to land.

What causes traffic, though? On a fundamental level, if you think about it, as long as there isn't an accident or something else, can't we all drive down a highway with only minor variations in speed? You would like to think that, but that's living in an ideal world (which we do not inhabit). Even if we have a scenario where cars are going down a highway at (mostly) the same speed, a small kick to the system in the form of a person cutting across lanes or the typical problem of people rubbernecking to see the results of an accident on the other side of the highway. If the people who are slowing down overbrake, then that will send a shock wave of traffic through the cars behind them. Even if there never was any actual accident or construction, this traffic slowdown can last for hours. We all have been stuck in a traffic jam with no apparent cause. This has many regional names, but it is known as traffic flow instabilities in the literature. I prefer "phantom traffic jams."

Another cause for traffic is just the density of traffic on the road. If all spots on the road are filled, and you try to add another car, there will be a slowdown. To explain this in a much more mathematical way, the speed that you can go is inversely proportional to the density of traffic. That is to say, as more cars enter the road, you will need to slow down. The actual relationship between these variables is described by a series of (partial?) differential equation and is very difficult to solve. One interesting consequence of these equations, though, occurs when you are stopped at a traffic light. Imagine that you are the fifth car back in the line and the light turns green. If everybody performs perfectly (obviously not a realistic situation), you will be able to start going as soon as the lead car going in the other direction passes your car. This is a bit of an idealized situation, but for human drivers, that's the way it would work.

There is another way (a better away, perhaps). What if we took humans out of the mix? What if we had self-driving cars? In the case of self-driving cars, they could be programmed to communicate with each other to all start accelerating at the same time when the light turns green. Who hasn't thought to themselves, "If everybody hits the gas at the same moment, and we all have uniform acceleration, we could all go at the same time. And then we all make it through that light"? Well, self-driving cars could do that. They would also eliminate people jumping across lanes unexpectedly, overbraking, tailgating, and people slowing down just to look at an accident. Self-driving cars could be the answer to traffic. Although accidents will not be fully eliminated, they will be significantly reduced as soon as human drivers are not in the mix. If only human drivers were outlawed from the roads, we could solve the vast majority of our traffic problems...

SPACE TRAVEL

SOLAR SAILS

Yoda: "The end for you, Count, this is."
Count Dooku: "...Not yet."
—Episode II

WHEN Episode II, escape from Geonosis

WHERE Geonosis

CHARACTERS Count Dooku

PHYSICS CONCEPTS Light, momentum

SHORT INTRODUCTION/BACKGROUND

Throughout the Star Wars movies there are many methods of transportation, but none as elegant as Count Dooku's solar sailer. The ship has a simple design of a pod with large sails attached. When traveling long distances through the depths of space, the solar sails are deployed and absorb sunlight to accelerate the ship. How practical is a solar sail? Could it allow for interstellar transport? Has this design ever been used? Does the design of Dooku's ship make sense?

BACKSTORY

Count Dooku is one of the primary antagonists in the second prequel. He heads the Separatist Alliance and is used by Darth Sidious to initiate the Clone Wars with the ultimate goal of forming a Galactic Empire.

To carry out this scheme, Dooku needs to travel to planets all over the galaxy, so he requires a solid long-range vehicle. If the solar sail is the solution to this problem, it is curious that we don't see more ships with this design throughout the movies. Maybe Count Dooku inherited wealth that allowed him to have a unique ship designed and built just for his purposes.

THE PHYSICS OF STAR WARS

A solar sail is a large sheet of reflective material that acts just like the sail on a sailboat, except that it relies on light for propulsion rather than wind. Light does not have mass, but it does have momentum. When light bounces off a surface, it gives a tiny push, much like wind bouncing off a sail imparts momentum to a ship. For light, only a tiny amount of momentum is transferred per photon, but stars emit a lot of photons. For instance, Earth's sun emits about 10^{45} photons per second, a number difficult to wrap your head around. In order to take advantage of all the photons from a star, the ideal solar sail would be perfectly reflective, lightweight, and as large as possible.

It is important to note that solar sails rely on photons and not solar wind. Solar wind refers to the matter a star emits constantly. It mostly consists of electrons and protons that were bouncing around inside the star and were expelled outward at high speeds. The amount of solar wind fluctuates but is always significantly less useful for propulsion than light is. On average, solar wind would provide about a thousand times less thrust than solar radiation. It is more useful in providing Earth with the northern and southern lights.

Let's first examine the shape and size of the solar sail on Dooku's vessel. Based on a comparison of ship size to the sail size, the sail looks about a hundred feet across the longer direction (major axis) and about seventy-five feet across in the shorter direction (minor axis). This gives an approximate area of just under six thousand square feet. This is a decent square footage if you are looking to buy a house, but it is tiny for an efficient solar sail. It will generate about one-hundredth of an ounce of force—equivalent to about two hundred feathers falling on you.

This probably sounds like a miniscule and useless amount of thrust. To put this thrust in perspective, sports cars can accelerate from zero to sixty miles per hour in a few seconds. This solar sail could accelerate the ship from zero to sixty mph in about seventy-two years. Even if Dooku used a substantially larger sail, he might be able to cut it down to five to ten years.

These calculations don't even take into account the fact that Dooku's solar sail is translucent. This means that there are precious photons passing right through the sail. It's difficult to tell from shots in the movie, but most still images portray the sail bent or arched like a parachute. The ideal solar sail would be completely flat so as to propel your ship forward. If you have a curve to your solar sail, some of the reflections would push your ship to the left and right, which is wasted thrust.

THE PHYSICS OF REAL LIFE

Solar sails have been in the minds of scientists all the way back to Johannes Kepler in the seventeenth century. Kepler observed that comets' tails point away from the sun, so he theorized there must be winds pushing the tail away. It wasn't until James Clerk Maxwell formalized the theory of electromagnetism that we first discovered that light itself has momentum. Jules Verne alluded to the idea of a solar sail in his book *From the Earth to the Moon*. From there it wasn't long before scientists began proposing spacecraft that used this momentum to achieve interstellar travel.

ELEMENTAL NAMES

Technically, Americans are incorrect calling aluminium, "aluminum." The IUPAC (the group of chemists who standardize nomenclature) wanted all scientists to agree on the names of elements. They decided that they would give aluminium to the British scientists and cesium and sulfur (as opposed to sulphur and caesium) to American scientists. Americans rejected the change, so the official designation is that element 13 is aluminium (although some scientists still call it aluminum).

In 2010 all of these theories were put into practice. The Japanese spacecraft IKAROS (Interplanetary Kite-craft Accelerated by Radiation of the Sun) was the first to use a solar sail for propulsion, although that was the secondary rather than primary means of thrust. IKAROS was designed not only to test the practicality of a solar sail but also to measure solar wind and gamma rays. Other craft have been designed, but most of the attempts have ended in either launch failure or failure to deploy the sail in space.

Since the ideal solar sail is highly reflective and lightweight, it is best if it's made out of a shiny metal with as little mass as possible. It is an added bonus if that metal also happens to be relatively abundant. The best candidate that we have found so far is aluminium. Most solar sail designs involve sheets of some kind of plastic that are about as thin as a bacteria cell. Those sheets of plastic are then coated with a layer of aluminium about as thick as the flu virus on one side. The sail still needs to be able to be folded up without breaking, creasing, or having any other irregularities, which could cause one side to have more momentum than the other.

The sail could be a number of different shapes, but the most commonly suggested is square. Calculations suggest that if we wanted a square solar sail that could lift a human, it would need sides about five miles long.

It may seem as if solar sails need to be impossibly large in order to be useful, but they just need to be large relative to their cargo. One proposed project is called Breakthrough Starshot, currently being funded by Yuri Milner and Mark Zuckerberg (among others). This project would send tiny spacecraft with comparatively large sails to Alpha Centauri, the nearest star system to Earth. These spacecraft would consist of little more than a camera, antenna, and sail. In order to propel them more quickly, they would be accelerated using the most powerful ground-based lasers for an initial kick. The goal is for the spacecraft to reach speeds around 20 percent of the speed of light so that they can make the journey in twenty to thirty years.

HYPERSPACE

WHEN All movies	
WHERE The galaxy	
CHARACTERS Most primary characters	
PHYSICS CONCEPTS Space-time, special relativity	

SHORT INTRODUCTION/BACKGROUND

In 1905 Einstein formalized his theory of special relativity. According to this theory, the speed of light is not only a constant, but also the universal speed limit. Theories proposing how one could travel faster than the speed of light have been put forth, ranging from wormholes to tesseracts to time travel. Some of these theories take advantage of additional dimensions that we cannot see. But how realistic are these suggestions? Are there more than three dimensions? Is it possible to travel faster than light? What is a wormhole and how would it allow us to travel great distances in a short amount of time?

BACKSTORY

In Star Wars, hyperspace is extra-dimensional space through which ships can travel so as to move across the galaxy faster than would be allowed by traveling through real space. In order to do this, a ship must be equipped with a hyperdrive. But going to hyperspace is not without its dangers. "Traveling through hyperspace ain't like dusting crops," as Han Solo explains. "Without precise calculations we could fly right through a star or bounce too close to a supernova." With such severe risks, it is important to rely upon hyperdrive computers.

THE PHYSICS OF STAR WARS

Hyperspace is, in theory, a set of extra dimensions beyond the three that we experience daily. These extra dimensions are able to connect distant points in real space. This allows for faster-than-light speeds (in a sense). For example, consider the flight from Tatooine to Alderaan. If Owen turned on a laser pointed directly at Alderaan (and we assume that there are no obstructions and the beam will stay accurately aimed enough to be detectable at Alderaan) at the same moment the *Millennium Falcon* jumped into hyperspace, the *Millennium Falcon* would arrive before the laser beam reached Alderaan. It seems as if the *Millennium Falcon* traveled "faster than light."

There are problems with this theoretical explanation. One is the idea that cause and effect rely on things happening in a particular order. More simply, for one event to cause a second event, the first event must happen before the second. That seems easy enough and unrelated to hyperspace, but the concept of simultaneity throws a wrench into everything.

Consider the following: you are sitting on a chair next to a high-speed railroad track, and you decide to launch two fireworks at the same time, one on either side. From your perspective, they launch at precisely the same moment. If your friend were to ride on a train traveling close to the speed of light as the fireworks were launched, that friend would see the fireworks launch at different times. An event that is simultaneous for you would not be simultaneous for your friend. Similarly, you could launch the fireworks at different times such that in your friend's reference frame they launch simultaneously. The catch is, if your friend's train were to travel faster than the speed of light, the order in which the fireworks launch will appear different to you (as a stationary observer) versus your friend (as an observer traveling faster than light).

You may think, well, fireworks are a silly example. Who cares if you disagree on the order in which the fireworks were launched? However, this thought experiment shows us the interrelationship between speed and the sequence of events. The laws of physics don't care what those events are. Imagine firing a blaster (event 1) and the bolt hitting the target (event 2). Or reading a book (event 1) and telling a friend about what you read (event 2). As you can see, the order in which these events happen would be nonsensical when reversed. Technically, it would be possible for the *Millenium Falcon* to fly faster than light past Alderaan as it explodes and arrive at the Death Star in time to stop the weapon from firing in the first place.

There are ways in which traveling through hyperspace would not require a violation of relativity, though. The idea that two points in real space are connected by a "tunnel" taking advantage of additional dimensions is not unheard of in physics theories. These connections between points in space-time are called wormholes.

Here's how a wormhole works: Hold a piece of paper in front of you and fold it in half. Now take a pencil (or other sharp object) and poke a hole through the folded paper. Now imagine that an ant wants to walk from one side of the paper to the other. If it walks along the surface of the paper, it will have to walk all the way up and around the fold. On the other hand, if the ant walks through the hole, it can get from one side of the paper to the other much faster. The ant itself never traveled faster; it just made it from one location to the other faster.

Whereas the paper is a two-dimensional surface, three-dimensional space as we understand it could be folded through a fourth dimension to create connections between two points. Because our minds have only ever experienced three-dimensional space, this is impossible to visualize fully. Still, if a hyperdrive were able to distort space-time such that it warped and created a hole between Tatooine and Alderaan, traveling through hyperspace would not violate any laws of physics. It would just require tremendous amounts of energy to accomplish these jumps.

THE PHYSICS OF REAL LIFE

This probably all sounds fantastical; something that couldn't happen in reality. As far as experimentally verified physics is concerned, that's true. There are theories, though, that indicate there could be additional dimensions of reality yet undiscovered. Perhaps the best-known example of this is string theory. At this time, there are five different formalizations of string theory, all of which cannot be falsified by current data. M-theory is a possible unification of all string theories according to which each individual string theory is a special example of the generalized M-theory.

The basic premise of all string theories is that everything in the universe is made up of tiny strings, which are either wrapped in a loop or exist in a straight line. Just as strings on a guitar oscillate in particular ways to make notes in a song, the strings making up the universe oscillate in different ways to create subatomic particles.

One of the other ideas of string theory is that there are more than the three spatial dimensions and one time dimension that we know. Depending on the specific formulation of string theory you are referencing, there are different proposed numbers of dimensions. For instance, in bosonic string theory, there are a proposed twenty-six dimensions.

So where are these extra dimensions? Why can we not see them or experience them? Like most things involved with physics on the border of human knowledge, we use analogies to describe the results. Imagine that you are an astronaut in the International

A NUMBERS PARADOX

Although going through the full mathematical derivation of bosonic string theory is beyond the scope of this book, the existence of these twenty-six dimensions relies on a seemingly incorrect mathematical result. If you were to take the sum of the counting numbers (1+2+3+4+5+...) all the way through infinity, the result of that addition is not infinity (as any sane person would suspect). In fact, mathematicians have proved that the sum of these numbers gives the baffling result of $-\frac{1}{12}$. The channel Numberphile on *YouTube* has videos in which mathematicians discuss proofs of this if you want more information.

SPACE TRAVEL

Space Station looking down at New York City. You will be able to see the grid of streets lit up at night. From your perspective, the streets will look like one-dimensional lines; things can go along them, but there is no width to go across them. Having been on a street, you know that you can walk across a street (not just go along it) and that you could even jump up and down while crossing the street, but from space you are too far away to see those details. Similarly, on our human-sized scale, we may be so far away from these compact dimensions that we cannot see the intricacies of them.

These dimensions are often described in terms of what is known as the Planck length. Some people suggest that this is the shortest possible length. The Planck length can be visualized in this way: Look at the width of a human hair. This is about a tenth of a millimeter across. If this hair were scaled up to be the size of the observable universe (about 10^{27} meters across), in the scaled-up version the Planck length would be the width of a human hair. Another way of saying this is that a human hair is about 10^{31} Planck lengths across. That is ten million times the number of stars in the observable universe.

SHIP DESIGN

WHEN All films	
WHERE The galaxy	
CHARACTERS All	
PHYSICS CONCEPTS Newton's laws, ionization, conservation of momentum	

SHORT INTRODUCTION/BACKGROUND

When it comes to having the coolest ship in the galaxy, design is crucial. When a couple of people are running for their lives, you don't want them to turn away from your ship because they think it's too much of a piece of garbage. But design of ships is also crucial for aerodynamics, battle-readiness, and general ability to perform the function set out for it. As far as the ships we see in Star Wars, do their ship designs make sense? How much do they really need engines if they have hyperdrives? What is an ion engine and how does it work?

BACKSTORY

Immediately after the opening crawl of *A New Hope*, we see the Imperial Star Destroyer Devastator pursuing *Tantive IV*. These are just the first of many different ships that we see in the Star Wars saga. From TIE fighters to the *Millennium Falcon*, X-wings to MC75 star cruisers, we see a huge diversity in shape and design. A few technologies seem to span the many ships, including hyperdrive cores, ion engines, and shields. How do the different designs indicate the uses of different ships? Do the designs of ships make sense? How can the *Millennium Falcon* make the sharp turns seen in *The Force Awakens*?

THE PHYSICS OF STAR WARS

Although they come in many shapes and sizes, most of the ships seem to be propelled by some variation of ion engines. An ion engine works like any other rocket booster in that it propels material backward that causes an equal and opposite reaction propelling the ship forward. Ions, being charged particles, can be accelerated to high speeds by applying extreme voltage. The problem with ion engines is that ions tend to be single atoms, and the amount of momentum transferred to the ship is proportional to the mass of the particle being expelled. For instance, a xenon atom (commonly used in NASA's ion engines) has a mass of about 10^{-25} kilograms. Even if it were shot out the back of the ship at the speed of light (the upper limit on its speed), it would take 32 kilograms of xenon to accelerate a Star Destroyer from zero to sixty miles per hour. It would take about a thousand years to produce that much fuel at current rates of xenon production.

This leads to the question: "What's the point of these engines anyway?" Obviously, they could be using a heavier element and have endless stores of it, but that doesn't make much of a difference in the calculation. Ion engines are best suited for ships that need to travel a long distance and can accelerate for years at a time. Given that hyperdrive technology exists, there is not much use for an ion thruster.

But what if these engines were not emitting ions? What else could they use to propel themselves. The engines appear blue, so what if the engines were just giant blue lasers? Continuous operation lasers can be up to megawatts in power. A megawatt laser emits about 10^{24} photons per second, which is still about a hundred times worse than an ion engine. There goes that idea.

Setting aside the seemingly incomprehensible engine design, what about a ship's ability to fly? In space, all you need to get somewhere is acceleration and time. In an atmosphere, the story is completely different. In the gravitational field of a planet or moon, a ship needs to produce lift to be able to stay aloft. Airplanes produce lift using wings, but the few ships in Star Wars that have wings seem to only have them as decoration (I'm looking at you, Kylo Ren). The clearest example of this is the TIE fighter's vertical "wings." These could not produce any lift in the traditional sense, so unless the engines were angled downward a bit when flying in atmosphere, a TIE fighter would slowly fall to the surface. The *Millennium Falcon*, on the other hand, seems to have a bit of an angle to it. Unfortunately for it, its underside is angled down (from nose to tail) while its topside is angled up. These would create counteracting down-force and lift.

The maneuverability of ships varies quite a bit. The X-wing, TIE fighter, and *Millennium Falcon* all seem to be very maneuverable ships. As we see in *The Empire Strikes Back*, Star Destroyers seem to have little to no maneuverability since they make no attempt to avoid colliding during the battle of Hoth. In order to create the maneuvers demonstrated by the smaller ships, there has to be some form of thruster exerting a force on the ship or a force generated by adjusting parts of the ship so that the air creates the force. With jet engines, the nozzle can be adjusted to redirect the exhaust to generate these forces. With a ship like the *Falcon*, though, it only (visibly) has the array of engines in the back of the ship. This would allow for thrust but would only offer limited maneuverability. X-wings and TIE fighters at least have wings, which (in principle) could be adjusted to provide the necessary forces (even though there is little to no evidence for this in the films themselves). This all may be moot because of the existence of repulsorlift technology (see Repulsorlifts). I find it difficult, though, to believe that these would be able to generate the maneuvers that we see in *The Force Awakens*.

THE PHYSICS OF REAL LIFE

As impractical as ion engines prove to be in the Star Wars universe, they are very useful for us. Since we do not have hyperdrive technology (yet), our best method for accelerating something to high speeds for interstellar travel is the ion engine. The *Dawn* spacecraft launched by NASA in 2007 used ion thrusters to move from Earth to Vesta to Ceres (dwarf planets in the asteroid belt). The ion engine on *Dawn* would require about four days to go from zero to sixty miles per hour, which means it would lose in a drag race. However, since it accelerates for years at a time, it can eventually reach a speed of about a mile per second. This is nothing record-breaking (the ISS is traveling about five miles per second), but this demonstrates the ability to accelerate a ship to exceptionally high speeds with little fuel.

Maneuverability of aircraft has long been an important aspect of aerial combat. In World War II dogfights, the pilots would need to outmaneuver each other in close-combat fighting. Since the Vietnam War, NASA, the US Air Force, and a number of other international groups have been trying to perfect what is known as supermaneuverability. This consists of maneuvers that would be

ROLL, PITCH, AND YAW

Airplane maneuvers are discussed in terms of their ability to induce rotations around each of the three axes. These rotations are referred to as roll, pitch, and yaw. Roll consists of rotations around an axis going from tip to tail of the aircraft (this is where the roll in "barrel roll" comes from). Pitch is a rotation around an axis that goes from wingtip to wingtip, causing the nose to point up or down. Yaw is rotation around an axis that goes from the top of the aircraft to bottom of the aircraft, allowing the nose to turn to the left or right.

impossible with just aerodynamics. New aircraft designs and techniques such as thrust vectoring were developed in order to make supermaneuverability a reality.

Thrust vectoring (an idea for which had been around for a long time) was proven to be capable of producing supermaneuverability with the Rockwell-MBB X-31 airplane. This experimental jet was a joint design of the US and German governments. Its engine had paddles in the back where the exhaust exited the engine. The paddles could be dynamically moved in order to direct the thrust up, down, left, or right in addition to its overall backward direction. Remember, whichever way the exhaust goes, the plane will go in the other direction. If the paddles direct the exhaust downward and to the left, the plane will have upward pitch and have positive yaw.

This supermaneuverability allowed for jets to perform maneuvers previously impossible. The X-31 was able to hover in place with the nose pointing nearly vertically or perform Pugachev's Cobra (a midair "stop" maneuver, which effectively points the nose of the aircraft straight up so that a pursuing aircraft will overshoot and the pursued becomes the pursuer). Perhaps with a skilled enough pilot the X-31 could pull off the maneuver made by Rey in *The Force Awakens*.

HAMMERHEAD SHIP

WHEN *Rogue One*, Battle of Scarif	
WHERE Low Scarif orbit	
CHARACTERS Kado Oquoné	
PHYSICS CONCEPTS Rotations, torque, moment of inertia	

SHORT INTRODUCTION/BACKGROUND

At this stage in history we have very few vehicles equipped to fly in space. It would be rare for two spaceships to collide with one another, although this is known to happen with satellites. In the Star Wars universe, spaceships are the most ubiquitous means of travel. Despite their ubiquity, the movies do not show many collisions between space vehicles. In *Rogue One* though, we see a Hammerhead Corvette push a Star Destroyer into another Star Destroyer. This scene looks like a problem straight out of an introductory physics textbook. Could a small ship push a larger ship so effectively? How would this all work? Did the film get it right?

BACKSTORY

During the climactic Battle of Scarif, the rebels are frantically trying to find a way to disable the shield so that the Death Star plans can be transmitted to the rebel fleet. In a self-sacrificial move, Kado Oquoné drives his Hammerhead Corvette into the side of a disabled Star Destroyer. This sets off a chain reaction of events leading to the destruction of the shield gate, which opens up the possibility of transmitting the Death Star plans. With the transmission of the plans, the rebels are given…a new hope… Too on the nose? Yeah, I thought so too.

THE PHYSICS OF STAR WARS

To a physicist, the scene of the Hammerhead Corvette running into a Star Destroyer can be described as a force applied to a rigid body causing rotation about its center of mass leading to a collision. Let's parse that sentence quickly. A force causing rotation is also known as a torque. The center of mass is a measure of (on average) where the bulk of the ship is located. Since Star Destroyers are wider in the back, we expect the center of mass to be toward the rear of the ship. The center of mass is important because if we apply a torque to a floating body, it will rotate about its center of mass. But how do we calculate the center of mass? How do we know the amount of torque?

In order to find the center of mass of a Star Destroyer, we have to know some parameters. There is no canon specification of the dimensions of the Star Destroyer, but we can use the non-canon number of about one mile in length. Given this dimension, you can measure the remaining dimensions of a Star Destroyer from images of one. Interestingly enough, if you assume a uniform mass density of an object, then the center of mass can be calculated without knowing the total mass of the object. Because the Star Destroyer is symmetric, we know that the center of mass lies along the line connecting the tip to the tail. Also, since we only want to know about rotations around a vertical axis, we don't need to worry about the vertical mass distribution. Using these simplifications,

THE AMOUNT OF MASS

Even though we don't care about the vertical distribution of mass, we do care that the ship becomes taller at the back as more of its mass will be weighted back there. To do this precisely is exceptionally complicated, but we can use some assumptions that are okay. In this calculation we assume that the mass density is approximately linearly increasing from tip to widest part and then constant from the widest part of the ship to the engines. Admittedly, this is certainly not precisely correct, but it will give us a reasonable enough answer.

we find the center of mass is located about 0.72 miles from the front end of the Star Destroyer.

Now that we know where the center of mass is, we can use it to learn about the torque applied by the Hammerhead Corvette. To do this we need to relate the torque, moment of inertia (how hard it is to rotate something), and its angular acceleration. The torque is the product of the force applied and the distance between where that force is applied and the axis of rotation. In order to check the reality of this scene, we will need to find the thrust produced by the Hammerhead Corvette and compare it to typical spaceship thrusts.

We can find the distance from the center of mass to where the Hammerhead Corvette is exerting its force (and thus the distance between the force applied and the axis of rotation). Using similar triangles and geometry, we can determine that the Hammerhead is about 0.44 miles away from the center of mass of the Star Destroyer. Now that we know the distance, we then need the moment of inertia to be able to determine the thrust generated by the Hammerhead. The moment of inertia requires a number of complicated integrals to find directly, but we can make a simplifying assumption. If we treat the front half of the ship as a point mass located at its center of mass and similarly for the back half of the ship, we can find the moment of inertia more easily. This, along with measuring the angular acceleration of the Star Destroyer from the scene, can tell us that the Hammerhead is generating about two hundred million pounds of force, which is about seventy times that of a space shuttle rocket booster. This number is within the realm of possibility.

THE PHYSICS OF REAL LIFE

At this time, we don't have spaceships designed for battle, much less specialized ones designed to ram another spaceship. What we do have is cars and boats, and sometimes boats ram boats or cars ram cars.

The case of boats ramming other boats sounds a lot more exciting than it really is. In order to ship materials via barges, tugboats push barges up and down rivers and canals. Barges are (often) unpowered rafts that can carry shipping containers. They are attached to one another, and a tugboat moves behind the line of barges and pushes it down the river or canal to its destination. During the industrial revolution this type of transportation of goods was accomplished with draft animals, which walked alongside the canals. How far we've come. Transportation of goods via barge has been declining in recent years but still makes up about 20 percent of goods transported in the US.

As for cars ramming other cars, perhaps the example most akin to the Hammerhead Corvettes from *Rogue One* is the PIT maneuver. The PIT (precision immobilization technique) maneuver is mostly used by law enforcement to stop a fleeing vehicle during a car chase.

The physics behind the PIT maneuver are similar to the Hammerhead Corvette. The police car applies a force to the tail end of the lead car. This force causes the rear tires to slide to the side (much like fishtailing if you have ever experienced that). Tires rely on the static friction between the tires and the road to move forward. By causing the tires to slide laterally, that static friction is lost, and the car loses its forward propulsion. Once the rear wheels lose traction, continuing to apply the force to the rear of the lead car causes it to spin around completely. If this were done in space, there would be no friction to worry about, and the police ship could redirect the other ship just as was demonstrated in *Rogue One*.

The PIT maneuver is similar to the "bump and run" from auto racing. In a bump and run, the trailing car will intentionally (or accidentally) bump the rear of the lead car. This can cause the rear tires to lift off the road and lose traction, so the lead driver must correct by slowing down. This allows the trailing car to accelerate around the car and take the lead. This technique (and the PIT maneuver) are highly dangerous at such high speeds and the bump and run is outlawed in certain types of racing. NASCAR's Car of Tomorrow tries to prevent accidental bump and runs by positioning the rear bumper lower than the front bumper. If there were a bump and run attempt, this would only apply additional downforce to the car in front, thus giving it more traction.

KESSEL RUN

"You've never heard of the *Millennium Falcon*? It's the ship that made the Kessel Run in less than twelve parsecs."
—Han Solo (Episode IV)

WHEN Episode IV	
WHERE Mos Eisley cantina	
CHARACTERS Han Solo, Chewbacca	
PHYSICS CONCEPTS General relativity	

SHORT INTRODUCTION/BACKGROUND

One of the most infamous quotes from the Star Wars movies is Han Solo's claim that the *Millennium Falcon* "made the Kessel Run in less than twelve parsecs." It is infamous because the parsec is a unit of distance rather than time. There have been all kinds of theories proposed to explain how this is possible. Maybe Han was able to spend more time in hyperspace than others (you only measure distance in real space). Maybe Han was able to fly close to black holes, thus shaving off distance from the run. Or maybe Han just talks a big game and he had no idea what he was saying and hoped that Obi-Wan and Luke didn't either.

BACKSTORY

In *A New Hope*, Obi-Wan and Luke need to get from Tatooine to Alderaan without the Empire finding them. They turn to smuggler

Han Solo and his copilot, Chewbacca. In an attempt to prove how good a smuggler he is, Han claims that he was able to make the Kessel Run in less than twelve parsecs. This apparently is convincing enough for the desperate Luke and Obi-Wan, so they agree to hire Han and Chewie. The Kessel Run referenced in this boast is a smuggler's route, named for the spice planet Kessel. A popular commodity, spice is often smuggled and sold on the black market.

THE PHYSICS OF STAR WARS

Let's consider how one could reasonably describe a flight speed between two points in terms of the distance traveled. Imagine two points on a sheet of paper. There is a definite shortest distance between those two points (namely, a straight line). Now imagine that an ant wants to walk from one point to the other and that the straight-line distance is ten feet. The ant walks one foot per second. It will take the ant ten seconds to walk that distance. Boring, right? Let's spice it up a bit.

What if directly in between our points, we put an antlion (a creature much like the sarlacc in that it stays in a hole and waits for ants to fall in its mouth)? That ant can no longer make it between the two points in ten seconds. Let's say the ant has to walk an extra three feet to make it around the antlion, still moving at one foot per second. This means the ant has to travel for thirteen seconds. A second ant now comes along and is more daring, so it inches closer to the antlion and does the walk in twelve seconds. The first ant, not to be outdone, does the walk again, but this time goes even closer to the antlion. Since both ants walk at the same speed, there is equivalency between comparing distances and times. In the case of most races that we think of, the vehicles are not traveling the same speed (otherwise podracing would be quite boring). If for some reason all ships on the Kessel Run were going the same speed, there would be an equivalency between distance and time. The ant analogy suggests that there is some hostile territory or other such obstacle between the smugglers and Kessel, thus making a direct route impractical.

Now let us consider a different scenario with the same two ants trying to walk between two points. This time the points will be fifteen feet apart, but the ants will still walk at one foot per second. Walking straight there, the ant will take fifteen seconds. But, what if there is a human with a twelve-foot stick that the ants can crawl onto. One ant walks from the starting point directly onto the stick, continues walking along the stick for the full twelve feet. The human moves the stick so that its other end is at the endpoint. The ant has managed to walk from one point to another while only traveling twelve feet. This is like hyperspace; there is an additional dimension you can travel through, making real space travel much faster than the laws of physics would typically allow.

Now suppose that instead of a stick or an antlion, there is a pit in the paper in between the starting and ending points. Imagine for a second that the paper is bent around this pit in such a way that it mimics the shape of water going down a drain. The direct route is no longer the shortest path between the two points. The direct line involves going down into the pit and then all the way back up. Going around the outside of the pit also could take a long time. The shortest path is somewhere in between. But imagine that the ant, relying on friction to stay on the paper (and not fall in the pit), is nervous about walking a section of paper that is too bent. The ant will then err on the side of the edge of the hole. Now a second, more daring ant comes along and walks along the curved path. The trip requires a shorter amount of distance than the previous ant's. This situation calls to mind the idea that there is some large gravitational anomaly (like a black hole or cluster of black holes) in between the smugglers' starting point and Kessel, bending space-time.

All of these could be explanations for why the Kessel Run would be best measured in distance.

THE PHYSICS OF REAL LIFE

Lots of people have criticized Han's claim, but we are all guilty of just about the same thing. Imagine you were talking to a neighbor who just moved to town. He asks, "How far away is the store?" and you answer "Oh, about fifteen minutes." Your neighbor now has a very good sense of how far away the store is, yet he asked a distance question, and you answered with a time. This is so commonplace that you probably don't even notice yourself doing it. We are able to communicate about distances in terms of time as long as we agree on the speed we are traveling. The ants in the earlier examples all went the same speed. In your conversation with your neighbor, the implication is that you are talking about fifteen minutes by car (or bicycle, hopefully). If you want to describe the distance from Washington, DC, to London, you might say that it is about seven (plane) hours.

At a certain point in physics classes, it becomes cumbersome to carry around so many constants, like the speed of light (c), Planck's constant (h), or Boltzmann constant (k_B). Wouldn't our equations be more elegant if these constants just went away? There is a system of units called natural units that accomplish this. It is not worth discussing all of the ways these units are defined, but one of the nice things about the system is that distances are measured in time. We agree on speed as the speed of light and then measure all distances in the time it takes light to travel those distances. This is where the distance unit "light-year" comes from. The nice thing about this is that the speed of light (c), then becomes one light-second per second, or 1. In this set of units, Einstein's famous equation is just $E = m$.

This may seem like a trick of numbers, but there is something more fundamentally intuitive with this set of units. Imagine you are teaching and want to test your class. When you grade the test, what scale do you use? Probably you would score people on a scale from 0–100 percent, or from 0 to 1. The highest score a student can receive is a 1, the lowest score a student can receive is a 0.

This is the same idea as natural units when it comes to speed. The fastest that any object can travel is the speed of light, so we assign that speed, 1. The slowest anything can go is nowhere, so we assign that speed, 0. Every other speed is somewhere in between. This gives us a better intuition for how fast something is really going.

I don't recommend adopting this in your everyday life. "Officer, I was only going 1.43×10^{-7} in a 9.67×10^{-8}."

HANDHELD WEAPONRY

BLASTERS

"Hokey religions and ancient weapons are no match for a good blaster at your side, kid."
—Han Solo (Episode IV)

WHEN Episode IV, opening battle	
WHERE Orbit of Tatooine	
CHARACTERS Stormtroopers	
PHYSICS CONCEPTS Plasmas, lasers	

SHORT INTRODUCTION/BACKGROUND

Blasters are the standard handheld weapon of the Star Wars galaxy. They look and sound cooler than the guns we're used to, but also seem to be much less accurate weapons (if stormtrooper performance is any indication). Sometimes they kill instantly, sometimes they are used to stun, and sometimes they miss the target completely. The closest camera angle on a blaster wound comes when Leia is hit in the shoulder on the forest moon of Endor. It seemed to burn her clothes and inflict a minor burn. So why did everyone in the galaxy start using them? How do they work? Do we have anything equivalent in our universe?

BACKSTORY

Blasters are ubiquitous in Star Wars. The Galactic Empire and the Rebel Alliance use them, droids use them, and smugglers and bounty hunters seem especially inclined to use them. To some (namely, Jedi) they are "clumsy and random," but to most they are an asset. In one especially controversial case, someone dodges a blaster shot from only a few feet away while seated. This is the "Han shot first" scene from Episode IV; in the original release, there is no need for Han to dodge a shot because he shoots and kills Greedo the bounty hunter preemptively. In later releases, the scene is edited so that Greedo shoots, Han dodges, and then shoots back. Knowing that shots can be dodged at such close range may help to explain the clumsy and random nature of the weapon.

THE PHYSICS OF STAR WARS

Some sources call the blaster a laser weapon and some refer to it as a plasma weapon; we will explore both options. For a description of how a plasma or laser actually works, see the following Physics of Real Life section.

If it's a plasma weapon, a blaster would compress tibanna gas, a substance mined in places such as Cloud City. After being compressed, the tibanna gas is energized and launched out of the barrel of the blaster toward its target in the form of a bolt. In this scenario, the blaster bolt is a beam of expelled plasma confined to a finite shape, often a line. We can look at some real-world materials to understand this since tibanna is a fictional substance.

First, we need to know at what temperature tibanna gas becomes tibanna plasma. The temperature at which materials turn into a plasma is fairly consistent, so we could estimate that a reasonable temperature at which tibanna gas becomes a plasma is 360,000°F. If such a gas came into contact with your body, it would transfer its heat to you. At very high temperatures, most materials have approximately the same specific heat (ability to store thermal energy). We can say that a plasma bolt at 360,000°F would most likely vaporize any part of your body that it hit.

There is a problem with blasters shooting plasma, though. A plasma is made up of a soup of charge particles that will experience forces from electromagnetic fields. A plasma bolt shot at seventy-three miles per hour (a decent estimate for the speed of blaster bolts in Star Wars) would only require a field about a million times weaker than Earth's magnetic field to cause the bolt to move one-and-a-half feet to the right or left (if the target is thirty-three feet away). This could explain why blasters are apparently random and stormtroopers appear to have terrible aim. The slightest bit of stray magnetic field could unexpectedly alter the path of the bolts. In fact, if a stormtrooper were firing a shot on Earth, the bolt would not only miss its target, but would travel in such a tight circle that it would hit the gun from which it was shot.

Given how much a stray magnetic field would affect the trajectory of a plasma bolt, maybe blasters are indeed laser guns as indicated in the original script. The accuracy of a laser gun is much higher, since light is more difficult to redirect. It also requires less energy to produce a bolt. When picturing a laser, you probably think of one that will not cause any harm or destroy instrument panels when you "shoot" them. This is because laser pointers are the most prevalent and are (mostly) class 1 lasers. A laser weapon would most likely be a class 4 laser, which can burn the skin, ignite combustibles, and definitely cause vision damage.

Typically, class 4 lasers are above 500mW of power, which would mean that if one was in contact with your skin for several seconds, it would cause severe burns. Higher-power lasers would obviously do more damage more quickly, but this seems consistent with the damage that Leia receives when she's hit on Endor.

Perhaps the best argument against the idea that these are laser beams is that all light travels at the speed of light. These blaster bolts are traveling significantly slower than light; they travel closer to a hundred feet per second rather than the 186,000 miles per second that light travels. In the movies it takes a second or two from when the blaster is fired and the person is hit. If this were an actual laser traveling at the speed of light, that would be the time it would take to hit somebody on the Moon while standing on Earth.

Neither of these explanations match what is seen in the movie. If we have to choose one explanation to be the most likely, it is the plasma explanation. It is more likely that there are no magnetic fields in scenes where blasters are fired than engineers who designed the blasters found a way to slow down light.

THE PHYSICS OF REAL LIFE

As noted before, plasmas are more of a soup of charged particles than atoms in the traditional sense. The nuclei and the electrons move around unbound to each other. In order to prevent the atoms from reforming, we have to keep the plasma at a high temperature. Temperature is a measure of the average translational kinetic energy of molecules—how quickly they're moving. As temperature increases, particles will move around more frantically. Eventually, atoms will jiggle enough that they break up into their constituent parts. Since those parts are moving quickly, they can no longer stick together. (Picture trying to grab somebody's hand while she drives past you on the highway; she's moving too fast to form that bond. Please don't try this at home.) At this point, the substance becomes plasma. For hydrogen, this happens at 288,000°F. The temperature threshold is higher, though not significantly, for larger atoms.

Since these particles are charged, they can be influenced by an electromagnetic field. Charged particles moving in such a field will travel in a circle. Test chambers at places such as CERN (the European Organization for Nuclear Research) employ this principle to keep plasmas contained.

The word *laser* is an acronym for "light amplification by the stimulated emission of radiation." Let's take this word by word. Obviously, lasers are a form of light, so that one is easy. Amplification references the fact that a laser beam starts as a single photon bouncing between two mirrors inside the device. As it bounces back and forth, it stimulates the emission of more photons just like it. The process of an atom emitting a photon is referred to as radiation. In short, a photon in a laser recruits a number of photon friends that are just like itself. Then they're emitted from the laser in the form of a beam.

But not every light is a laser. So how does a laser work? Lasers are made with a chamber with reflective walls full of particular gasses (most common red lasers have helium and neon in them). Lasers require this gas to be in a state called population inversion.

This name refers to the number of atoms in an excited state (higher energy) at any given time. Most atoms in a gas at room temperature are not excited, but a few are. If we invert these populations (so that most are excited and only a few are not), we achieve, wait for it, population inversion. This is achieved by applying a voltage to the gas from either a battery or an electrical outlet.

When we reference an "excited atom," what we really mean is that one of the electrons in that atom has extra energy that it would like to give off. If a photon (light particle) with the proper wavelength (color) passes this excited atom, the atom will give off a matching photon. The battery then will excite the atom back to where it started so the process can repeat itself. Do this times a million (actually more like times 10^{23}), and you can produce a coherent beam of light.

LIGHTSABERS

WHEN All films	
WHERE The galaxy	
CHARACTERS All Jedi and Sith	
PHYSICS CONCEPTS Plasmas	

SHORT INTRODUCTION/BACKGROUND

The most iconic weapon of the Star Wars universe is the lightsaber. Lightsabers combine the sophistication of a sword with the fantastical feeling of a laser. They make Jedi feel both ancient and futuristic. Lightsabers have been described as laser swords (i.e., light) or potentially contained plasma. Either way, they are formidable weapons capable of deflecting blasters, deftly dueling, and cauterizing wounds. Are they even possible, though? If so, would they be able to absorb or deflect blaster bolts? Wouldn't a lightsaber just pass through another lightsaber in a fight? Do we have any devices that have similar properties to lightsabers even if they aren't the real thing?

BACKSTORY

Lightsabers are what makes Star Wars, Star Wars. On the surface, they're just fun to watch. They also help us feel the conflict and emotional upheaval the characters are experiencing. What would the iconic "I am your father" moment in *The Empire Strikes Back* be without the preceding lightsaber duel between Luke and Darth Vader? They are clearly a brilliant element of the films, but...does the science hold up?

The extended universe of Star Wars establishes that lightsabers are powered (and colored) by kyber crystals found in locations around the galaxy (including Jedha from *Rogue One*). Do these crystals have any basis in reality? Putting that aside, are all the different colors and designs practical?

THE PHYSICS OF STAR WARS

Lightsabers are usually around three feet long. How easy it is to create a three-foot beam depends on whether it is a beam of light or a beam of plasma.

Beams of light are tricky to contain because photons move at the speed of light; thus, they are very difficult to turn around or stop midair. Perhaps the easiest way to create a three-foot-long beam would be a mirror opposite the hilt of the sword to reflect the light. This is obviously not the design presented, since when they're off, lightsabers are no larger than their hilts. The sound of a lightsaber turning on could be the sound of a mirror extending outward as if it were uncapping a container full of light, but there are still other issues.

For example, the fact that the beam is visible light (we can definitely see it!). If you've ever shone a laser pointer on your arm, you know that it won't slice skin. The power of a visible light laser pointer would need to be upped by a factor of about a million before it could do any damage, and a laser of that power would require an extensive cooling system. Further, as far as we know a beam of light, no matter how powerful, is incapable of deflecting a bolt of plasma shot by a blaster. Similarly, a beam of light could not absorb plasma.

If we consider the beam to be plasma, there is a different set of concerns. A well-designed electromagnetic field could, in principle, contain a plasma to a size of about three feet (maybe by sending the plasma in a highly elliptical path to create roughly the shape of a cylinder). Plasmas are also hot enough to cauterize wounds and melt metal (both aspects of lightsabers seen in the movies). We're off to a good start, but problems arise if we consider dueling plasmas. Expecting some free-floating plasma to clash with some other free-floating plasma is like...expecting soup to clash with other soup. The two plasmas would actually be attracted to each other (as they are made up of charged particles) and become one. This would also make it difficult to deflect a blaster bolt, but it could explain how it is able to absorb force lightning.

As mentioned in the Ion Cannons section, the color of plasmas depends on the temperature. In that respect, a red lightsaber would be lower energy than a green one. This would also be true if they were made out of light since green light has more energy than red light. To generate plasmas of red or green is quite challenging. Most plasmas, both in labs and in stars, are generated predominantly using hydrogen. This means we know the color of hydrogen-based plasmas quite well. If we made a cobalt plasma, though, would it appear to be a different color? We will just have to do that experiment to find out.

Plasmas are hot, and being close to a plasma would also be hot. Since plasmas are often at temperatures of millions of degrees, holding a plasma stick in your hand would lead to some severe burns. The sun is ninety-three million miles away, and we need to wear sunscreen to protect us from it—despite the fact that there is an atmosphere blocking most of its harmful radiation. Holding a miniature stick-shaped sun in our hands would require at least SPF 10,000.

There could certainly be some other explanation as to how lightsabers work, but it would either be not based in reality (e.g., using the magic of kyber crystals) or an incredible feat of engineering involving much more than just light or even plasma.

THE PHYSICS OF REAL LIFE

How close are we to making a lightsaber? As discussed in the Blasters section, contained plasmas exist in science labs today. At this time these plasma containers are the size of rooms, but so were computers sixty years ago. Now we have pocket-sized computers way more powerful than those room-sized computers, so that means we'll have pocket-sized plasmas in sixty years, right? Probably not.

In order to contain a plasma, we need superconducting magnets that need to be cooled down to close to absolute zero in order to operate. In order to cool down something that low, we need to surround it with gallons of liquid helium (which is about −450°F). No matter how advanced our technology gets, gallons of liquid will still be larger than our pockets. If we could develop high-temperature superconductors, though, that would be a completely different story.

One interesting advance in the realm of lightsabers comes from the contained beam of light design. In 2013 scientists at Harvard and MIT worked together to create what has been described as a "photonic molecule." They sort of convinced two photons to move together in tandem, much like two atoms bound together in a molecule. They accomplished this by sending the photons into a super-cooled gas of rubidium atoms. At this stage, though, it is only two photons, which is far away from a self-contained beam, and it too relies on temperatures close to absolute zero.

Although it would be quite an engineering challenge, one way to create a lightsaber at home would be a combination of current sword technology with a technology developed by Burton, Inc. Burton has developed a way to generate 3-D images that appear to levitate. They use a high-powered laser to concentrate a beam (of nonvisible light) on a particular point in air. The beam heats up the air to the point where it becomes a plasma, causing it to glow. Burton uses these points as pixels in low-resolution images. In principle it would be possible to create a device (or use an array of multiple devices) to generate a straight line of these points, thus creating a beam of plasma. It might be tricky for the laser to heat up points in the air fast enough for a smooth swinging action to occur, though. If this isn't living up to your childhood dreams, join the club.

A lightsaber-esque device is already being used regularly via electrosurgical instruments. Electrosurgery is a process where an electronic device is heated via electrical current to the point where it can cauterize wounds. In order to be able to do this, the device needs to reach temperatures of hundreds of degrees Fahrenheit. This may not exactly be a lightsaber, but it performs one of its functions.

SEISMIC CHARGES

"Hang on son, we'll move into the asteroid field. And we'll have a couple of surprises for him..."
—Jango Fett (Episode II)

WHEN Episode II, orbit around Geonosis	
WHERE Geonosis asteroid field	
CHARACTERS Jango Fett, Boba Fett, Obi-Wan Kenobi	
PHYSICS CONCEPTS Sound, light, vacuum pressure	

SHORT INTRODUCTION/BACKGROUND
In space no one can hear you scream. Space is a vacuum, and sound requires a medium (stuff to move through) in order to propagate.

Without air (or some other material), sound cannot exist. So, where is the border between where sound can exist and where it can't? Where is the border between the atmosphere and space? Scientists generally consider this to be sixty-two miles above the surface of Earth. At this point, air is about 100,000 times less dense than at sea level; thus you need a particle density around 10^{13} per cubic foot for sound propagation. So, without air, could these seismic charges work? Would other sound weapons like concussion grenades work?

BACKSTORY

A common critique of the scientific accuracy of the Star Wars movies is the presence of sounds (most often explosions) in space. Many are happy to excuse this issue on the grounds that the films just wouldn't be as fun without the sound of explosions. There is one instance that might be more difficult to overlook, though. In *Attack of the Clones* the bounty hunter Jango Fett uses weapons called seismic charges to try to evade Obi-Wan Kenobi, who is tracking him. A seismic charge creates a blast wave of sound that causes destruction in a disk surrounding it. When Jango deploys the seismic charges, he and Obi-Wan are navigating through an asteroid field. Could a weapon like this be used in space? Would it destroy asteroids and spaceships alike?

THE PHYSICS OF STAR WARS

A seismic charge is supposed to be a sound bomb—a blast of sound energy that expands into a disk shape. So, what is this energy exploding outward? The name seismic charge calls to mind an earthquake. Seismic waves travel through the crust of a planet. This means that on some level, any bomb that shakes the surface of a planet is "seismic" as long as it explodes close to the surface. Technically speaking, a seismic charge uses this type of wave to create destruction (akin to a concussion grenade, which is designed to do damage with the shock wave emitted from it).

A seismic charge makes sense when deployed near a planet's surface, but what about in an asteroid field? Without a medium for the explosion to move through, it can't generate a sound blast. Such a blast pushes everything away (think air and small debris) from the epicenter, leaving a vacuum. As the blast expands and loses energy, the outward force causes a collapse back into the center. This type of explosion relies on a medium (such as air) to offer a counterpressure to the explosion. If this blast were to hurt you, it would be from a bunch of tiny air (or dust) particles slamming into you at high velocity. Without the medium, there is nothing to propel at high speeds, and therefore no blast.

Could a seismic explosion work in space? Warning: it's going to be a stretch. The critical issue is the lack of a substance for the sound wave to travel through, so we need to know what types of matter could be in the area. The "asteroid field" Obi-Wan and Jango are navigating is actually the rocky rings of the planet Geonosis, so we have some options. First, each asteroid could have its own atmosphere. Second, the asteroid field/rings could be inside the range of the atmosphere of Geonosis. Lastly, the asteroids themselves could have enough matter to propagate sound waves.

It is highly unlikely that each asteroid has its own atmosphere. The atmosphere of Earth's moon is 10^{13} times less dense than that of Earth, and sound does not exist on our moon. If a body the size of Earth's moon cannot maintain enough atmosphere for sound waves, none of the rocks around Geonosis would be able to do so.

It is similarly unlikely that the rings of Geonosis are within its atmosphere. If its rings are at all similar to Saturn's, the innermost rings would be around 4,350 miles above the surface of Geonosis. As mentioned earlier, the borderline between space and the atmosphere is about sixty-two miles above the surface of Earth. Even if Geonosis's atmosphere extended 10 times farther out, it would still be a few thousand miles short of the rings. Rings are generally outside the range of a planet's atmosphere because if they were within the atmosphere, they would burn up in the same way a meteor does.

The last option is the most feasible. Most particles in the rings of Saturn are just microscopic dust. If the density of dust particles in the rings of Geonosis were high enough, sound could pass through them and allow the seismic charges to be effective. Further, the first charge detonated would create additional tiny dust particles, making the later charges even more effective. But, even this can't easily explain a seismic explosion. For instance, let us assume Geonosis's rings had the same mass as Saturn's rings (3×10^{19} kg). Even if we spread out the atoms of the rings as much as possible, they would still not be dense enough by a factor of about ten thousand.

One other aspect of the seismic explosions shown in Episode II is their unique shape. The explosion starts with a sphere of energy, which has what appears to be jets of energy going out from the poles. The sphere eventually pancakes out into an expanding disk. The astute observer may recognize that this shape is very similar to the orbitals of an electron in a hydrogen atom. This is characteristic of any spherically symmetric object rotating around a particular axis, but how the angular momentum for such rotation is generated is a bit ambiguous.

Since the direction of the angular momentum is crucial to hitting your target (since the angular momentum decides which way the disk expands outward), it would make sense to have that be a controllable aspect of this weapon. The simplest way to control the angular momentum of such a device is to have spinning plates (called gyroscopes) inside of the bomb prior to its detonation. By controlling which way those plates are spinning, the direction of the explosion can be sent toward your target.

THE PHYSICS OF REAL LIFE

One aspect of the explosion of seismic charges we haven't addressed is the fact that the explosion was visible in the film. In order to see something, it has to emit (or reflect) light. How does sound emit light? Sound can produce light via a process called sonoluminescence. If you imagine a bullet being shot into water, the bullet leaves behind a trail. That trail is (in some respects) a bubble left behind the bullet. As the bullet pushes the water out of its way, it leaves behind a vacuum (called cavitation). Eventually, that vacuum sucks the water back in on itself and then collapses. When a bubble collapses on itself, it creates high internal pressure. When the bubble reaches maximum compression, the energy inside the bubble can be strong enough to emit light. The precise mechanism by which this happens is still being debated, but it has been demonstrated. Devices called bubble chambers use this principle to measure everything from neutrinos to dark matter.

Another interesting consequence of a shock wave passing through air is cloud formation. Air can only hold a certain amount of water vapor without condensing into a cloud; how much water vapor the air can hold mostly depends on the temperature and pressure of the air. When a compression wave passes through humid air, the pressure momentarily increases significantly and then decreases. This spike in pressure causes the water vapor in the region of high pressure to coalesce into a cloud. This can be seen along the shock wave of explosions or around jets traveling at supersonic speeds. Certain types of explosions can create different kinds of clouds. For example, a large explosion (for example, from a nuclear bomb) causes mushroom clouds. When the explosion first happens, it creates a ball of hot gas (essentially a fireball). Much like a hot-air balloon, that ball will rise in the air. As it rises it leaves behind an area of low pressure. That low-pressure area sucks surrounding air and dust into it. This is what forms the stem of the mushroom. As the fireball moves up, it flattens out, causing the cap of the mushroom.

It may seem absurd that a sound wave could kill, but they can definitely be very damaging. As mentioned earlier, concussion grenades use this principle. The most likely injury is to the eardrums, which will rupture with extreme blast waves. The next most

KRAKATOA

The largest known shock wave on Earth occurred when the volcano Krakatoa erupted in 1883. Although Krakatoa is in Indonesia, the explosion itself could be heard distinctly as far away as Sydney, Australia. Some instruments measured the shock wave as many as seven times because it left Krakatoa in all directions (radially). This meant the shock wave circumnavigated the globe three and a half times over several days (the wave was traveling about 675 miles per hour, so it took about thirty-six hours to make it around the globe once). Sailors' eardrums forty miles away were ruptured.

vulnerable organ is the lungs (or other hollow organs such as the intestines). When the pressure wave reaches its maximum pressure, it can cause the lungs to explode (if the wave is strong enough). The negative pressure sucks the air out of you, which can also be extremely traumatic. After the shock wave comes an intense wind that can carry with it dust particles moving quickly enough to tear the skin. If you are able to survive this, there is often secondary damage such as brain trauma (imagine being hit by a hundred football players at once) or damage from being thrown against another object.

GRENADES

WHEN Episode I, Battle of Naboo; Episode VI, opening scene

WHERE Naboo; Jabba's palace

CHARACTERS Princess Leia Organa, Jabba the Hutt, the Gungan army

PHYSICS CONCEPTS Explosions, plasma

SHORT INTRODUCTION/BACKGROUND

Grenades were developed to allow strikes from a distance without requiring precision. As baseball legend Frank Robinson said, "Close only counts in horseshoes and hand grenades." Bombs accomplish the same goal but with increased range and destructive power. How do grenades work? What are the different varieties of hand grenades? Will we ever be able to harness the power of a nuclear bomb in a handheld weapon?

BACKSTORY

The Star Wars movies depict multiple handheld explosive devices. We are introduced to the small handheld "thermal detonator" in Jabba's palace in Episode VI when Leia (disguised as bounty hunter Boushh) threatens Jabba with one. Thermal detonators can supposedly cause a fusion reaction in a contained radius of about twenty feet. In Episode I, the Gungan army uses balls of plasma called boomas (of varying sizes, including some that are handheld) against the droid army. Upon detonation, boomas produce a small range of electrical discharge, which destroys or disables droids and other machinery. Are these weapons feasible? Are they depicted realistically?

THE PHYSICS OF STAR WARS

Thermal detonators simply aren't realistic. Fusion reactions are already used in bombs, but the reactions can't be contained in a radius of twenty feet. In order to initiate a fusion reaction, materials such as hydrogen, helium, or lithium must be forced close enough together that their nuclei fuse. In Star Wars these materials are replaced with fictional materials (such as baradium), but the principle still stands.

Typically, fusion weapons have a core of material surrounded by a ring of explosives. In order to produce enough energy to force the materials close enough together (both by heating the materials and increasing the pressure surrounding them), kilotons or even megatons of TNT are necessary. This will lead to a blast radius of greater than twenty feet; such a radius would result from about 0.13 tons of TNT, but an explosive would not produce fusion—though it does make the detonator sound more futuristic.

What about boomas? Are they more probable in design or effect? Boomas appear to be spheres of plasma contained by a field of some sort (so that they are safe for the user). When the containment field is ruptured, the plasma causes a small electrical storm on the target, thus disabling it. Boomas come in various sizes, from a few inches across for the handheld variety to about ten feet across for use by heavy artillery.

Containment of plasmas has been discussed extensively elsewhere (see the Frozen Blaster Bolt and Lightsabers sections), as has the damage done by plasma (see Blasters). Here, let's just focus on the design of a containment field that would not only hold the plasma but protect the user and also be able to rupture and release the plasma onto the target. Can we contain plasmas in a popable bubble?

Because plasmas are a soup of charged particles in motion, they will generate a magnetic field. For instance, the magnetic field of Earth's sun is around 1 Gauss (except in active regions where it's about a thousand times larger). Proportionally, a booma would have a field around a billion times weaker. As such it would probably not be able to exert a strong enough force to retain a protective sheath, but there are magnetic fluids (called ferrofluids) that, if the field is strong enough, could provide the protective sheath. Given the behavior of ferrofluids, though, the sheath would probably be spikey like a hedgehog.

There are a lot of engineering challenges to this idea. We would need to make the ferrofluid orbit the plasma inside so that it does not get sucked into the plasma and destroy the booma before its intended use. On top of that, we would need to find a way to thermally insulate the ferrofluid from the plasma. Plasmas are very hot, and our ferrofluid would probably either evaporate or lose its magnetic properties from being heated too much.

THE PHYSICS OF REAL LIFE

There are many types of grenades. Fragmentation grenades and concussion grenades are two types. Concussion grenades are discussed in the Seismic Charges section; fragmentation grenades are designed to do damage by sending pieces of shrapnel flying in all

MASS TO POWER

To determine how much mass we need for a given amount of energy, we need to calculate the cleverly named "Q value" of a reaction (named because the variable Q was chosen to represent it). The Q value of the reaction is an application of Einstein's famous formula $E = mc^2$. The Q value of a reaction is the amount of energy released by a conversion of mass to energy, or said in equation form, $Q = (m_i - m_f) c^2$. The difference in masses is called the mass deficit. When the Q value is positive, we say the reaction is exothermic (gives off energy) and when the Q value is negative, we say that it's endothermic (takes energy to happen). Converting a Ping-Pong ball's mass to energy in this way would be equivalent to a 43-kiloton explosion (larger than the explosion over Nagasaki in 1945).

directions. These do not have nearly the same power that a hand-held nuclear device would have—if one existed.

The closest we've gotten was one prototype that would have used W54 warheads and weighed about fifty pounds. They would have been launched from a stationary gun called the Davy Crockett and had a range of a mile or two. These warheads are on the lower limit of practicality for producing a nuclear explosion. The blast radius of something like this would be about 170 feet; the radius of lethal radiation was about a quarter mile.

In order to understand the power of a nuclear weapon, it is first important to establish an energy scale. Most contexts use the scale of the explosion of one metric ton (1,000 kilograms) of TNT. That's about twice the energy released by the same amount of gunpowder and about six times less than the amount of energy by the explosion of the same weight of gasoline. If we use one ton of TNT as our reference point, it corresponds to about 4.184×10^9 joules per ton. In order to produce this amount of energy from fission, we only need about 46 billionths of a kilogram of material. To put this into perspective, there is sixty-four times this amount of mass in a snowflake.

The idea of a magnetic liquid known as a ferrofluid deserves a deeper dive, though. These are typically created by having nano-magnets (very tiny magnets) in a fluid solvent (such as water). These fluids respond to magnetic fields and allow for the visualization for magnetic fields in a 3-D way unlike the traditional iron filings, which only allow for a 2-D representation of the field. Although these are mostly cool to look at (for the average person, you should watch a *YouTube* video of one if you have the chance), there are some very interesting applications. Ferrofluids are used in hard drives to prevent dust accumulation, friction reducers for magnetic pieces, and can even act as cooling agents for loudspeakers.

ELECTROSTAFF

WHEN Episode III, rescue of Chancellor Palpatine
WHERE Bridge of the *Invisible Hand*
CHARACTERS MagnaGuards, Obi-Wan Kenobi, Anakin Skywalker, General Grievous, Chancellor Palpatine
PHYSICS CONCEPTS Electricity

SHORT INTRODUCTION/BACKGROUND

Staffs are frequently associated with elderly men and/or mystical powers (think Gandalf in *The Lord of the Rings* or the Staff of Ra in *Raiders of the Lost Ark*). Staffs feature heavily in the mythologies of many gods. Because of the association with magical powers, in a lot of pop culture we also see staffs used as weapons. In the real world, staffs are more associated with hiking than weaponry, but obviously the idea of hitting someone with a big stick has been around for a long time.

BACKSTORY

Star Wars provides a weaponized staff known as the electrostaff. Predominantly used by General Grievous's personal guards, the electrostaff consists of a six-foot stick with sustained electricity surrounding the last foot or so of either end. We see them used with moderate effectiveness against Obi-Wan and Anakin as they rescue the chancellor from General Grievous in Episode III. How difficult would it be to have a staff with electrified ends? Would there be any issues with wielding a weapon such as this? Would it be able to stop the blade of a lightsaber? If thrown hard enough, would one of these be able to break the window of a spaceship?

THE PHYSICS OF STAR WARS

A large electric potential is required to create sustained electrical discharges over a distance of about a foot. In order to generate just a spark across that distance, you need to create a potential difference large enough to ionize air. On Earth, that means about a million volts per foot. That sounds like a lot, but the design of such a weapon would be easy enough. If each end had a metal ring about a foot from the edge and a high-voltage electrode on each end of the staff, it would act like a capacitor continuously charged by an internal power source and then discharged via the electrical breakdown of the air.

So how does this all work? There are two metal rings, one at the end of the staff charged up to a very high voltage. The other ring, closer to the center of the staff is grounded. This creates what is called a capacitor, a device designed to store electrical charge. As the charge on a capacitor increases, the electric field between the two rings increases proportionally. Eventually, the electric field between the rings reaches a point where it can separate the electrons from their atoms and briefly turn the air into a highly conductive plasma. Once charge is allowed to flow between the rings, they become fully discharged (because the negative charge on one moves to cancel out the positive charge on the other). It is then up to the power source to recharge these metal rings again.

The creation of this weapon is feasible, but that doesn't mean it'd be practical to use. The problem with an electrostaff is that you are charging up the ends, and the most convenient place to discharge them is the metal rings (one foot away from the ends). If you put the end of the staff less than a foot from any metal surface, it would likely discharge there instead. Try watching any of the fights between Obi-Wan and one of the MagnaGuards and see how often the ends of the staff are within a foot of something metal. As much as it's a generally good idea to keep the ends of your weapon away from your body, it is especially imperative when you are made out of metal and your weapon will fry your circuitry.

Would one of these staves be able to stop a lightsaber or crack through the window of a spaceship? The short answers are no and if thrown hard enough, respectively. One could potentially stop a lightsaber, but not in the way shown in the movies. In order to create the lightning at the ends of the staffs, there must be a large electric field. Since a plasma (see the Lightsabers section) is a soup of charged particles, the staff's electric field would exert a strong force on all of the charged particles and could disperse the beam of a lightsaber (if it isn't held in place by a containment shield). As for breaking a window, the strongest glass will break with around 1GPa of pressure (around one tenth of what is needed to form diamonds). This means that a staff would need to exert a force of about two million pounds in order to crack the window on the *Invisible Hand*. The fact that the ends are charged does not increase the force, so we're basically wondering if a generic staff can crack a window, and the answer is...sure, if you throw it hard enough.

THE PHYSICS OF REAL LIFE

The physics of a staff is not very new or complicated. Since a staff is like a lever, it is capable of taking a small force applied by the hands and turning it into a large force applied to the target.

Devices like this have existed for about a hundred years now. The most dramatic example of a device that can cause electric arcs like the electrostaff is the Tesla coil. Named for Serbian inventor Nikola Tesla, the Tesla coil generates a high voltage, which can cause sparks to jump from the coil to the surroundings. Originally designing it as a possible method of wireless transmission of power, Tesla built a Tesla coil in Colorado Springs capable of generating

RATIO OF FORCE TO HIT

Roughly speaking, the ratio of the force applied by the attacker to the force of the hit on the target is equal to the ratio of the distances of the ends from the axis of rotation as described by

$$\frac{F_{attacker}}{F_{target}} = \frac{r_{target}}{r_{attacker}}.$$

THE PROBLEM WITH CHICKEN WIRE

Faraday cages can cause problems for you in your everyday life. If your house is old enough to have chicken wire holding plaster walls together, you may notice that your cell phone and Wi-Fi signals don't travel between rooms well. Chicken wire is a metal grid with openings small enough that radio and microwave signals (like Wi-Fi and cell phones) cannot get through easily.

lightning bolts over one hundred feet long. Today, Tesla coils are used for educational or entertainment purposes. One group, called ArcAttack, performs educational shows with musical Tesla coils, which play songs with their electrical discharge.

How would you win a fight against a foe who was employing either an electrostaff or Tesla coil as a weapon? You would need to avoid the electrical discharge (especially near your heart or other internal organs), since a few million volts would probably stop your heart and cause serious internal burns (see Force Lightning for more on the biological effects of electricity). Probably the simplest defense is a technology that predates human-controlled electricity: chain mail.

This may sound as smart as playing golf in a lightning storm, but it would actually keep you relatively safe. Electricity is lazy: it will take the easiest path to the ground. Human bodies are fairly conductive when compared to air, trees, and other common surroundings. The conductivity of metal, however, is much higher than your body, thus making the chain mail more attractive to the electricity than your body. When lightning (or a Tesla coil or an electrostaff) shocks a person wearing chain mail, the electric current won't bother with going through the human because it can reach the ground via the mail. Of course, with sustained electrical discharge through the mail, it could heat up and cause burns, but hopefully the occupant of the mail could win the fight quickly enough that this wouldn't be a problem.

This is why the safest place in an electrical storm is inside your car. Not because the rubber tires discourage lightning strikes (the small gap between the bottom of your car and the ground is nothing to lightning), but because the metal frame will conduct the lightning safely to the ground without hitting you. This is true in airplanes as well. A metal container (such as a car or a plane) that neutralizes the effects of electric fields is called a Faraday cage, after scientist Michael Faraday.

HEAVY WEAPONRY

ION CANNONS

WHEN Episode V, Battle of Hoth	
WHERE The Hoth system	
CHARACTERS Rebel and Empire forces	
PHYSICS CONCEPTS Heating	

SHORT INTRODUCTION/BACKGROUND

Ion cannons are referenced as heavy weaponry on both sides of the Rebellion. There are ion cannons guarding the secret rebel base on Hoth. There are also ion cannons on Star Destroyers. There are tradeoffs for these more powerful weapons. They are able to take down a Star Destroyer in one shot, but their firing rate is much slower than that of a smaller laser weapon. How would an ion cannon work and what is the bolt actually made of? Would we have to use different ammunition for the bolt to be red versus green? Could an ion cannon vaporize an asteroid? How much energy would it take to operate a weapon like this?

BACKSTORY

In the opening of *The Empire Strikes Back*, the secret Hoth base is discovered by the Empire. In the ensuing evacuation, the rebels use their ion cannons to cover evacuating transport ships. With a couple of shots, they are able to take down a Star Destroyer. Later, as the *Millennium Falcon* is being pursued by the Death Squadron, Han and company fly into the Hoth asteroid field. During the pursuit, a Star Destroyer uses its cannons to vaporize asteroids to try to minimize damage dealt to the ship. In a single blast the asteroid is blown into microscopic pieces.

THE PHYSICS OF STAR WARS

The destructive power of ion cannons is only explicitly shown once. This is at the beginning of *The Empire Strikes Back* when a Star

EMP

An electromagnetic pulse (EMP) is a short burst of electromagnetic energy. This can come from things as common as lightning or things as terrible as nuclear weapons. Electromagnetic pulses are capable of disrupting or completely destroying electronic devices.

Destroyer is taken out by a few blasts from the ground-based ion cannons near the rebel base. The blasts do not seem to do much structural damage, but they appear to send a strong enough electrical current through the ship to fry all of its computers. This would be the same effect as a very strong electromagnetic pulse. A blast of this strength would probably require about the same energy that a US household uses in one year.

A second example of a heavy weapon in use is when the Star Destroyer vaporizes an asteroid. Although this is not explicitly shown to be an ion cannon, it is as powerful as one. In order to vaporize something, it needs to be heated to the point where it melts and then evaporates. Estimating how much energy this would require requires knowing the precise size and makeup of the asteroids in the Hoth field. Typical asteroids in the solar system are predominantly iron, so we can use the properties of iron in our estimate. To estimate the size, we can look at the size of the impact of an asteroid colliding with the underside of the Star Destroyer. Putting all of these pieces together, we can say that the blast from the heavy weapons on the Star Destroyer would be about 10^{14} J, or about ten times the amount of energy released in the detonation of the atomic bomb over Hiroshima.

It is clear that it would require large amounts of energy to power these weapons, but it is not impossible to accomplish this. There are other concerns, though, when it comes to firing high-powered weapons such as this. For instance, a beam of ions can undergo a process known as blooming. If all the ions in the beam have the same charge (say an electron beam), they will repel each other over time causing the beam to spread out and become ineffective when it reaches its target. Thermal blooming also occurs when ions run into particles in the air. The fact that it's snowing on Hoth will only increase the amount of blooming that will occur.

There are other concerns for both the ground-based ion weapons and potentially those mounted on the Star Destroyer. When firing a beam of ions in a magnetic field (which admittedly Hoth does not need to have), the ions will experience a force perpendicular to the direction of motion. This will lead to the particles moving in a circular path (see the Blasters section for more).

Even if Hoth does not have a magnetic field, certainly Star Destroyers fly through regions close to planets or stars that do have magnetic fields.

If one was to design an ion weapon, a design that is either disk-shaped or spherical would make sense. In order to heat up the ions enough to be an effective weapon, it would be easiest to have the ions move in a circular path while being accelerated. Once you wanted to fire, the magnetic field holding them in this path could be turned off and the weapon would fire off a beam in a straight line. This could explain why it would take a while between shots as it takes time to accelerate the beam of ions to sufficient speeds as well as the spherical shape of the Hoth-based ion cannons.

THE PHYSICS OF REAL LIFE

As of now, there are not any known ion cannons in operation. Of course, there could be some developed in secret that have never been used. But the effects depicted in the films are still capable of being generated.

Electromagnetic pulses (EMP) have been developed as weapons. An EMP is generated with every detonation of a nuclear weapon, but they have been weaponized outside of nuclear bombs as well. One of the earliest recorded effects of an EMP came from the Starfish Prime nuclear test in 1962. In this test conducted by the United States, a 1.44-megaton bomb was detonated in the atmosphere. About nine hundred miles away in Hawaii, hundreds of streetlights went out and burglar alarms were triggered. Had this device been detonated over the continental US, the EMP likely would have been about five times more powerful, causing fires in power plants, destruction of electrical equipment, or excessively high currents through long power lines.

The EMP from a nuclear blast comes in three waves known as E1, E2, and E3. The E1 blast, the first and most intense, comes from the gamma radiation produced during the blast. It will fry electronics by inducing a voltage above that which electronics can handle (called the breakdown voltage). After E1 comes E2, which produces an effect equivalent to that of being struck by lightning. In general, lightning strikes don't affect our electronics because we have surge protectors in our homes (and similar devices on a much larger scale for our infrastructure). However, the E2 blast immediately follows the E1 blast, which very likely just fried all of our safeguards against the E2 blast. Lastly comes an E3 wave, which is a distortion in Earth's magnetic field rather than a direct effect of the blast itself. The effects from the E3 wave is similar to the effects from a magnetic storm on the Sun hitting Earth.

One last quick note is on the color of ion beams. An ion weapon would be the same thing as a plasma weapon since a plasma is a soup of charged particles (ions). Plasmas have characteristic colors that depend on their temperature. We can see this in the color of stars in our universe. Hotter stars appear bluer whereas colder stars appear more red. Stars that appear red are brightest in the infrared range, and the concentrations of other wavelengths of visible light are small compared to the amount of red. This means that a red plasma is very possible. As for green plasmas, though, we don't see these in stars. A star whose peak wavelength is in the green range has a high concentration of the surrounding colors (as green is in the center of the visible range), so that star will appear white (this is what our star happens to be). It is possible that a plasma created using something other than hydrogen (which is what stars rely upon) would have different optical properties.

DEATH STAR

"This station is now the ultimate power in the universe!"
—Admiral Motti (Episode IV)

WHEN Episode IV, Battle of Yavin	
WHERE Yavin	
CHARACTERS Luke Skywalker	
PHYSICS CONCEPTS Energy	

SHORT INTRODUCTION/BACKGROUND

The Death Star is one of the most iconic weapons in modern film. With the power to destroy an entire planet in a single shot, the Death Star was designed as a tool of intimidation as well as destruction. Obviously, a weapon (or space station) of this size has never been created, but would it be possible? How much energy would it take to destroy a planet? Would a planet explode in the dramatic way that we see Alderaan, Scarif, or Jedha did? What would the physical consequences of making such a large weapon be? How close have we come to creating weapons of this power? Are there natural events that match the destructive power of the Death Star?

BACKSTORY

The Death Star was in the works long before the Galactic Empire even existed. The first demonstration of its full power comes via the dramatic explosion of Princess Leia's home planet, Alderaan, in Episode IV. By the end of the same movie, the station has traveled a large distance across the galaxy in order to target the rebels' base on Yavin 4. With all the planning that went into building this massive station capable of interstellar travel, it is completely destroyed by a relatively small force of rebel pilots. Is this reasonable? We are given more of a backstory on its vulnerabilities in *Rogue One*, but what about its overall design?

THE PHYSICS OF STAR WARS

There is a lot of interesting physics to look at when it comes to the Death Star. Perhaps the most relevant question is whether a laser weapon could cause a planet to explode and, if so, by what mechanism?

For instance, the explosion of Alderaan in Episode IV is depicted as though there is a big pile of dynamite in the planet's core, which the Death Star ignites. For an explosion like this (involving combustion) to occur, there must be heat (which a laser provides), oxygen (which the atmosphere of Alderaan probably has), and some form of fuel (such as oil). So maybe the Death Star's laser is melting through the crust of a planet and "igniting" a reservoir of oil and natural gas? The rapid heating of a large reservoir of fossil fuels mixing with oxygen in the atmosphere would cause a planet-wide explosion. However, there are a few caveats.

Caveat number one: the atmosphere of Alderaan must have oxygen. Since Leia (and plenty of other people) lived there, it is

---

HOW HOT?

To give a sense of how underpowered our laser technology currently is, the most powerful laser (which can only be turned on for fractions of a nanosecond) would need to be turned on for a billion years before it would be able to destroy the Earth.

HEAVY WEAPONRY

not unreasonable to assume that there is oxygen in the atmosphere, but it's a fictional place, so who knows.

Caveat number two: Alderaan must have had biomass compressed by the planet over millions of years into fossil fuels that were not mined in large quantities. This is a much larger assumption, but there are a few indications that it's a possibility. The planet appears very green and blue, so it probably had lush vegetation on it (also a potential indicator for photosynthesis and thus an oxygen-rich atmosphere). We don't have much evidence one way or another as to how much fossil fuel extraction might have been occurring on Alderaan. In any case, again, this all relies on the vegetation existing millions of years prior to the events of Star Wars. It also depends on Alderaan having reservoirs of fossil fuels when it was targeted by the Death Star.

Such a weapon's design seems awfully shortsighted for the Empire, since many planets in the galaxy probably don't have the needed fossil fuel inside them. It could be that the Death Star just heats the target to the point of destruction. In *Rogue One* we see the Death Star being used at partial strength. When fired in this way, it seems to create an effect more like a very large meteorite impact. When the Death Star destroys the holy city of NiJedha, the impact from the blast knocks debris into the upper atmosphere and nearly back up to the Death Star itself. If we assume that the impact is as catastrophic as the most destructive impact in recorded history (Fragment G of Comet Shoemaker–Levy 9 hitting Jupiter), then the Death Star would need to fire around 3×10^{45} photons, which would correspond to about 1.12×10^{27} J of energy. This is about the amount of energy put out by the Sun in three seconds or about the amount of energy used by humans over the span of around three million years (assuming the consumption rate of 2013).

So, what would really happen if you shot a giant laser into a planet? Well, if the laser were able to provide more energy than the planet can radiate away, the planet will heat up. After a few degrees of heating, the climate will be unsuitable for life. After a few hundred degrees of heating, the planet will start to melt. If the laser continues to heat the planet, it will heat to the point where it turns into a miniature star.

One last interesting consideration is the idea of trying to destroy a gas giant. Although direct measurements of the core of planets such as Jupiter have not been made, we have theories of what the core could be like. We suspect it has liquid metallic hydrogen. This is a state of hydrogen that does not exist naturally on earth and behaves like a liquid metal in both its thermal conductivity and its reflectiveness. Because it is highly reflective, it could potentially act as a giant mirror and reflect the Death Star's weapon back at itself. I would love to see the look on Grand Moff Tarkin's face as that happened.

THE PHYSICS OF REAL LIFE

What are the limits associated with building a Death Star–like weapon in the real world? The bottom line is that it would require amounts of energy well beyond our current capabilities. First, there is the simple limitation of the ability to gather the raw materials and construct them into a space station.

It is reported that the first Death Star had a radius around seventy kilometers. If we assume that the Death Star is made out of steel and is about 40 percent structure and 60 percent open space, then the mass of the Death Star is about 4.45×10^{18} kg (or about 16,500 times less massive than the moon). If we wanted to lift this much steel into space so that it was orbiting Earth as a second moon (at the same radius as the Moon's orbit), it would require about 2.78×10^{27} J of energy (about twice the amount of energy that goes into destroying NiJedha!). Further, we would have to bring air up to the Death Star. If we assume that the 60 percent open space is filled with air, that's another 6.58×10^{22} J (or about 170 years of 2013 energy consumption).

As an additional comparison, the International Space Station has a mass of about 400,000 kilograms and is at an orbit about 220 miles above Earth's surface. In order to accomplish this all in one go, it would require at least 1.5×10^{13} J of energy.

We rarely think about such things. Although people try to be more environmentally conscious by purchasing hybrid and electric cars, they typically don't discuss the fact that about half of the emissions associated with the life of a car are released before it is ever driven. This is not to say that electric or hybrid cars are bad; I'm just pointing out that if a person is concerned with the environment, she must consider the entire life cycle of the products she buys. This could mean using an old car longer, so as not to increase the demand for new cars. This may not be as dramatic as shooting an ion torpedo into an exhaust port, but it is a more real way to save a planet from destruction.

Next, we are limited in our planet-destroying-weaponry capabilities. Although humans have created weapons of mass destruction, none of them compares to the destructive power of the Death Star. The most energetic nuclear test in history was of the Tsar Bomba in 1961. This bomb released 210×10^{15} J of energy (three thousand times more energy than the Hiroshima bomb). The primary difference between the Tsar Bomba and the bomb dropped on Hiroshima is that the Tsar Bomba was a fusion bomb as well as a fission bomb. A fission bomb requires the decay of uranium and plutonium to create a runaway nuclear reaction, which gives up significant amounts of energy. A fusion bomb not only does this, but it uses the initial detonation to smash together smaller atoms (usually hydrogen, helium, or lithium) so that they generate energy from fusion. Fusion can produce a lot more energy than fission, but it requires extreme temperatures. The blast wave from such a bomb would flatten everything within fifteen miles, and the thermal damage would extend about seventy-two miles. If Chicago were hit with this bomb, the southern suburbs of Milwaukee would catch fire.

The destructive power of a nuclear weapon is devastating, but it pales in comparison to the destructive power of an impact with a comet or some other space object. This has happened to Earth before, as evidenced by the Chicxulub crater in Mexico. The crater is about 110 miles across and was probably created by an object about 6 miles across. The thought is that this impact led to the end of the dinosaurs (and over 90 percent of species on Earth at the time). NASA is tracking all the objects that cross Earth's orbit and are at least 164 feet across, so hopefully nothing this large will catch us off guard, but even an impact with an object this small could do significant damage. It might have been more energy-efficient for the Empire to design a giant space catapult that could mimic these kinds of impacts.

STARKILLER BASE

"We're not sure how to describe a weapon of this scale."
—Snap (Episode VII)

WHEN Episode VII, assault on the Galactic Senate	
WHERE Starkiller Base	
CHARACTERS New Order and the Resistance	
PHYSICS CONCEPTS Plasma, dark energy	

SHORT INTRODUCTION/BACKGROUND

The most impressive weapon in the Star Wars galaxy (at least so far) is Starkiller Base. Constructed inside the core of a planet, this weapon is able to shoot plasma projectiles across the galaxy and destroy entire star systems in one firing. Obviously, nothing this destructive has ever been created in our universe. Is it even possible to create a plasma weapon on this scale? Would a high-speed plasma projectile be able to destroy multiple planets as shown in *The Force Awakens*? How can the energy be stored safely inside the planet prior to firing?

DARK ENERGY

Dark energy is a theoretical type of energy used to explain why the universe is expanding. As of the writing of this book, it has not been experimentally verified, and we don't know much about it. It is referred to as "dark" because it literally is; it does not interact with light (we do have fairly good experimental evidence for this). Dark energy is believed to be a property of space itself, so it would provide nearly limitless amounts of energy if it could be harnessed.

BACKSTORY

Thirty years after the destruction of the second Death Star, the First Order emerged as a splinter group looking to restore the power of the dark side. After Palpatine was defeated and Darth Vader reconciled with his son, the republic was restored and a new senate was created. We don't know all the details yet, but we know the First Order wants to reestablish an empire and spread the influence of the dark side. To that end, they constructed a superweapon that would outperform the Death Star in every way. It is powered by sucking the life out of a nearby star, then firing plasma projectiles through hyperspace to targets across the galaxy. Draining a star, if possible, does seem as if it would create more destructive power than a laser beam, but it raises a lot of other questions.

THE PHYSICS OF STAR WARS

According to the novelization of *The Force Awakens*, Starkiller Base not only takes in the power of a nearby star, it also taps into dark energy (specifically a form of dark energy known as quintessence; more on that later) as a source of power. The base converts the star's power to another type of dark energy known as phantom energy. This is held in place by the planet's magnetic field, then fired through hyperspace.

This description of how the base works is impossible. First of all, if the base were using dark energy as a power source and was firing dark-energy bolts, it would not need to harvest energy from a star in order to fire its weapon. Another issue is that bolts made of dark energy would not appear as bright red streaks across the sky since it doesn't interact with light. Further, it is unlikely that a planet's magnetic field would be able to contain dark energy as it most likely does not interact magnetically. Lastly, if the projectiles are moving through hyperspace, it would be impossible for anyone in regular space to see them dramatically streak across the sky.

There is a much simpler way that Starkiller Base could operate. Unlike dark energy, plasmas do interact with electromagnetic fields. Stars are made of plasma, and it would be possible to construct a very strong electromagnetic field that would draw the plasma away from the star and into the core of the planet. Once the plasma was harvested, it would be possible to contain it using strong enough electromagnets. It could even be reasonable to call a device with such strong electromagnetic fields a thermal oscillator (this is the name given to such a device in the movie) since you would most likely need to have the plasma moving at high speeds in a circle. Once you wanted to fire the weapon, all you would have to do is stop containing the plasma when it was traveling directly at the opening in the base.

Technically, these projectiles don't need to travel through hyperspace, but it's probably more practical. The bolts would travel in a straight line with whatever velocity they left the weapon unless acted upon by another electromagnetic field. It is unlikely that the path between Starkiller Base and its targets would be unobstructed, plus the path might pass near enough to a neighboring star to deflect the plasma. This is not to mention that the target planet would probably have its own magnetic field. It is possible to take these deflections into account, but it would make the aiming systems significantly (and unnecessarily) more complicated.

THE PHYSICS OF REAL LIFE

Is it possible to create such a base? Realistically, no; we are centuries away from such technology, but there are some aspects that we could implement. For instance, it would be possible to hollow out the center of Earth. The reservoir of plasma for Starkiller Base would be housed by the planet the base is built into. There would be some severe dangers in doing this, but it is possible.

One of the first consequences would depend on what you did with all the dirt and rock you harvest. If you pile it all up, then it would effectively redistribute Earth's mass. Taken to an extreme, this could create a change in the local pull of gravity. Currently, the largest digging machine is the Bagger 293, which can move about eight million cubic feet of dirt per day (or about the volume of the *Hindenburg*). Using this machine, you would need to dig for about thirty years before the pile of dirt would be large enough to cause a noticeable gravitational anomaly.

If you could just magically make the dirt disappear, you could also affect gravity on Earth. For instance, if you were able to make the dirt vanish from inside Earth rather than just move it, the force of gravity on everything would decrease. Interestingly, though, it is impossible to measure the difference in gravity between a solid planet and a hollow planet which have the same mass. If Earth were a hollow shell of dirt that had the same mass as the current Earth, it would produce the exact same gravity for all the people on the surface as it currently does. The primary difference between a hollow-shell Earth and the actual Earth is that if you were able to get inside the hollow shell, you would be completely weightless and be able to float around. Being inside a shell with uniform density, all parts on it would pull on you in such a way to perfectly balance out leaving you with no sensation of gravity.

This may start to sound like a fun idea, but it doesn't take into account the most dangerous consequence of hollowing out Earth. At the center of Earth is a large quantity of molten iron sloshing around inside the core. This iron is essential to the survival of all life on the planet: it produces a magnetic field that surrounds Earth. This field is weak, about 100 times weaker than a refrigerator magnet. Despite its small magnitude, it redirects the near constant stream of high energy–charged particles that shower Earth at the North Pole and the South Pole. As these particles enter the atmosphere, they run into the air molecules, which give off light in the form of the Aurora Borealis and Aurora Australis. If these particles were not stopped, they would cause massive damage to the DNA of all living creatures and lead to the end of life on planet Earth. As a small consolation for the death of all living creatures, there would be the aurora at all latitudes at all times.

In some senses, the Large Hadron Collider (LHC), the main testing facility at CERN, was constructed in much the same way as Starkiller Base. A large ring was dug out from the Earth around which high-energy plasmas are sent. Because of this design, one could almost call the LHC a thermal oscillator. To be called a thermal oscillator, it should in principle have a very hot material (hence the thermal part) that moves back and forth. Although circular motion is not the same as oscillations, they are very related concepts in physics.

IMPERIAL WALKERS
(AT-AT/AT-ST/AT-ACT)

"We have spotted Imperial walkers!"
—Trench officer (Episode V)

WHEN Episode V, Battle of Hoth

WHERE Hoth, rebel base

CHARACTERS Rebel forces, Luke Skywalker, Imperial forces

PHYSICS CONCEPTS Center of mass, friction

SHORT INTRODUCTION/BACKGROUND

One of the most intimidating forms of ground weaponry employed by the Empire is the Imperial walker. These take a number of different forms, including the small all-terrain scout transport (AT-ST), the medium all-terrain armored transport (AT-AT), and the large all-terrain armored cargo transport (AT-ACT). They act as both weapons and as a way of transporting ground troops into battle. Is it possible to have a walking, hulking transport? Would it even make sense to create such a machine? What is the current status of our troop transports?

BACKSTORY

When the Empire discovers that the rebel base is on the icy planet Hoth, Darth Vader sends in his ground troops to destroy the base's shield generators. The first view of the AT-ATs is foreboding, but rebel pilots quickly learn that their legs provide an exploitable vulnerability. AT-STs are used to assault the forest moon of Endor, and AT-ACTs appear in *Rogue One* during the Battle of Scarif. Each time, the transports are fairly easily dispatched, calling into question their design and utility, especially when weighed against the practicality of getting them to the planets in the first place.

THE PHYSICS OF STAR WARS

Imperial walkers are a powerful symbol of the Empire, even if not the most effective weaponry in the galaxy. The larger AT-ATs are quite powerful and effective at fighting anything directly in front of them, but at the expense of their flanks. The AT-STs are best as scouts and nimbler guards of the rear. It seems as if the design engineers for the Imperial walkers prioritized intimidating their opponents over having a practical troop transport.

How does something so heavy maintain stability? For AT-ATs, the solution is to only lift a single foot from the ground at a time. In order for a transport like this to be stable, its center of mass must reside somewhere in the shape created by its supports. So, the center of mass of an AT-AT is always over a triangle formed by three feet on the ground. Contrast this method of transportation with that of a horse which sometimes lifts all four legs from the ground at once (but maintains stability through a combination of other factors).

This discussion of the center of mass of the vehicle needing to be over the top of the points of contact in order to remain stable raises the question: how are the AT-STs stable? If you watch the AT-STs walk, they sway back and forth. When they sway, they are

---

RUNNING HORSES

The story of how scientists determined that horses left the ground entirely is interesting in its own right. For centuries people were unsure whether a horse would fully leave the ground when at full speed. This was settled in 1878 when photographer Eadweard Muybridge took a series of twenty-four photographs of a mare named Sallie Gardner while she galloped. This series of photographs became known as *Sallie Gardner at a Gallop* and was shown at an exhibition in 1880, making it the earliest recorded instance of a motion picture being projected.

---

moving their center of mass back and forth to sit atop of the foot that is still on the ground. The feet of an AT-ST are just over seven feet apart; as they walk, the center of mass sways around one and a half feet to either side. This may not sound like much, but it is just enough to keep the center of mass over the planted foot.

Being stable does not, of course, inherently protect you from attack. AT-ATs are easily taken down if their legs can be wrapped in cables. After collapsing, presumably the shields go down as well, as we see a snowspeeder's blasters blow one to pieces. During the battle on Endor the Ewoks release two tree trunks to smash an AT-ST in its cockpit. Since those logs were based off the American redwoods, we can estimate their density to be about 450kg/m^3. They are probably about ten feet across and about thirty-three feet long, which means that the force they would exert on the sides of the AT-ST would be about 900,000 pounds (or about one-third of the thrust from the space shuttle during launch). This amount of force would almost certainly crush the cockpit of the AT-ST since it only takes 10 percent of this force to crush a steel drum.

THE PHYSICS OF REAL LIFE

As it stands right now, we do have troop transports, but none of them could be called a "walker." In fact, designing any kind of mechanical walking machine (whether it be a robot or a transport vehicle) has proven to be an engineering challenge in its own right. There have been robots (such as Honda's ASIMO) that are capable of walking, but only on very flat surfaces. A company called Boston Dynamics is designing robots of all different kinds. They have some that are designed to mimic dogs, which, as of a few years ago, were able to stay upright in snow and mud. Other robots are designed to handle all terrain, but these don't have feet; rather, they have rotating blades instead. None of these robots are close to being as top-heavy and stable as the Imperial walkers.

So, if we can't bring our troops into battle using an intimidating, walking behemoth, what do we use? Currently, the US army employs some troop transports that use wheels and some that use tracks to propel themselves. Although designs have mostly settled into these forms, there have been some unique failures along the way. Perhaps the Star Wars Empire is just in the midst of an awkward time in their design timeline.

Perhaps the closest thing to an actual Imperial walker was the cybernetic walking machine developed by General Electric in 1965. It was designed to be a quadrupedal hydraulic walking machine intended to help carry heavy equipment over rough terrain. Sadly, it was only about fifteen feet tall.

Other attempts to create all-terrain vehicles have had mixed success. One proposed design for navigating rough terrain was a vehicle propelled by giant corkscrews. These screw-propelled vehicles were able to plow through slush, marsh, and water reasonably well. However, they couldn't move on powdery snow, sand, or well-paved roads.

To propel a vehicle (or anything for that matter) forward, a force needs to be applied to the ground in the direction opposite to the desired direction of motion. The friction between the vehicle and the ground creates a forward-acting force on the vehicle moving it along. If there is not enough friction, the vehicle won't be able to propel itself. When you place a giant metal screw on top of a paved road, the contact area between the screw and the road is incapable of producing enough friction. The design relies on the screws being able to sink somewhat into the ground and grip onto it.

A last bit of hilariously bad design of a military vehicle is the Tsar tank. It was designed to look like a penny-farthing bicycle (the one with the giant front wheel) except the engineers felt that one giant wheel was not enough. The Tsar tank has two giant front wheels (about twenty-seven feet high) and a single rear wheel. In principle, this design could be very intimidating, but like the Imperial walkers, it failed in practicality. In order to keep the tank balanced, the rear wheel needed to support a lot of its weight, which meant that it sank into mud quite easily.

HEAVY WEAPONRY

SHIELDS

"Sanitation? Then how do you know how to disable the shields?"
—Han Solo (Episode VII)

WHEN Episode VI, Battle of Endor	
WHERE Endor	
CHARACTERS Han Solo, Princess Leia Organa	
PHYSICS CONCEPTS Energy	

SHORT INTRODUCTION/BACKGROUND
Shields date back to prehistory, so it's safe to say that they are some of the oldest technology that we have. There is quite a difference between the wooden shields of our ancestors and the energy shields depicted in the Star Wars films, but the principle remains the same. There is some sort of threat that needs to be redirected or absorbed. How do energy shields work? Would shields like this be visible? Are permeable shields like the Gungans have or impermeable shields like the ones on Scarif more realistic? Have we ever created energy shields?

BACKSTORY
The primary defensive technology shown in the Star Wars galaxy is the energy shield. These range from small shields that protect droidekas or Gungans to much larger shields that protect star destroyers, the Death Star, or even Starkiller Base. Some seem to just absorb blaster bolts, while others are deflective and cause shots taken at them to rebound into space. In almost all the Star Wars films a shield features heavily in the plot, yet very few design details are ever disclosed. What do we know about the shields that protected the Gungans, made the Empire send AT-ATs to Hoth, and forced Han to pull off a daring jump out of hyperspace near Starkiller?

THE PHYSICS OF STAR WARS

Since there are so many different shields featured in the Star Wars saga, let's discuss their general properties before going into the specific variations. Again, shields in Star Wars either deflect or absorb blasters, perhaps with some form of electromagnetic energy that isn't always visible. They also all seem to have some maximum amount of damage they can take before "going down." Let's look at the Gungan shields, deflector shields on ships, and the shields for Starkiller Base.

The Gungan shield, which we first see in *The Phantom Menace*, is the most intriguing to me. The shields are impervious to projectiles launched from artillery whereas droids seem to be able to walk right through them. The shield was most likely designed to have this strange property because it appears the Gungan weapons are better against infantry than against mechanical artillery. Their own artillery's range is completely inside the protective layer of the shield, so it does not offer them the same disadvantage. This shield is thus able to provide a protective barrier for their infantry to take on other infantry while opposing artillery is unable to affect the outcome of the battle.

Another interesting property of the Gungan shield is that it seems to be a bit elastic; ripples will move through it when it is hit with artillery fire or a droid passes through it. In order to have ripples in a surface like that, it needs to have some form of surface tension. To my knowledge, there is no form of electromagnetic surface tension.

The Gungan shield could work if it were a big dome of a shear-thickening fluid (non-Newtonian). What this means is that when you apply a force to the fluid, it becomes more viscous relative to the amount of force you apply. The most common example of this is the mixture of cornstarch and water. Ketchup has the opposite property—it becomes thinner with shear, which is why you have to hit the neck of the bottle in order to get the ketchup out. If the Gungan shield is a shear-thickening fluid, then high-velocity projectiles would be stopped by it because of the excessive force they would exert on the shield. A droid walking through the shield would apply a much smaller force and therefore could pass through it without issue.

Ships from X-wings to Star Destroyers all seem to have deflector shields protecting them. These shields are different from the Gungan shields in a number of ways. First of all, they are invisible to the naked eye. Also, they seem to cause the blaster bolt to explode close enough to the ship to still cause some form of response. It seems as if the shields are as strong as the power source you provide them. Larger ships have more powerful shields (presumably because they are able to generate more power). If you are a smaller ship trying to run away from a larger ship, you can divert all the power to the rear shields. If you can stay in front of the enemy, then you do not have to shield any other part of your ship. Once the shield has failed, though, all shots that hit will do damage to the ship.

Finally, there's the shield protecting Starkiller Base. It is said to have a "fractional refresh rate" which prevents anything traveling slower than light from making it through. Although it is unclear precisely what this means, one possibility is that the First Order has decided to power their shields with AC power (outlet) rather than DC power (batteries). DC power provides a constant voltage and constant current. AC power oscillates from positive voltage to negative voltage. This means that for a fraction of a second, there is zero current (and potentially no shields). Most places on earth use AC power in wall outlets, and there is a division of opinion over what the frequency should be. Very roughly, if you are in the Western Hemisphere (or Japan, the Koreas, or Taiwan), your power runs at a frequency of 60 Hz whereas everywhere else runs at 50 Hz. For most purposes, including for the Starkiller Base shield design engineers, this on/off oscillation has no noticeable effect on anything. But if the shields were indeed oscillating with a frequency of, say, 50 Hz, then it would potentially be possible for a ship to fly through the shields while they are momentarily off.

This poses an interesting physics problem. There is a tiny fraction of a second that the shields will be effectively off. The *Millennium Falcon* must completely pass through the shield during this interval. If we know the length of the *Millenium Falcon* and how fast it's going, we can figure out how long the shields are effectively off. The *Falcon* has a few advantages, though. As you move faster and faster, according to the theory of relativity the length of your ship will be shorter (something called length contraction). If we assume that the rebels were rounding when they said the ship had to be going light speed, and instead the *Falcon* only had to go 0.999 times the speed of light (0.999c), then the ship would be about twenty times shorter than its length when stationary. For something traveling at 0.999c, it would take about five nanoseconds for the *Falcon* to travel the necessary 1.5 meters. By this math, that means the shield can be penetrated by a ship as long as its strength is less than 10^{-7} of its full strength.

THE PHYSICS OF REAL LIFE

As anybody who has watched a documentary about warfare probably is aware, we do not currently have giant blue shields that can surround our troops to prevent them from being hit. Kevlar is our low-tech equivalent of shields. But what is our current closest equivalent to the electromagnetic shields in Star Wars? The best that we have at this stage is a Faraday cage discussed in the Electrostaff section.

What about that bit with the *Millennium Falcon* using length contraction to slip through the shields. Is that really possible? The ideas introduced by the theory of relativity can be a bit complicated and mind-bending, but just about every experiment ever designed to test relativity has shown that this is indeed how the world works. The fundamental idea is that the speed of light is a constant whether you are sitting still or moving. At face value, this seems easy enough to accept, but the implications of it are a bit strange. If you throw a ball 100 miles per hour while standing next to a tree, the ball will move away from the tree at 100 miles per hour. If you get in a car and drive 100 miles per hour and at the moment you pass the tree, you throw the ball 100 miles per hour in the direction the car is moving, the ball will move away from that tree at 200 mph.

This seems straightforward enough, but it is not the case with light. If you stand next to that tree and turn on a light, it moves away from the tree at the speed of light (186,000 miles per second). If you get in a car and turn on the light as you pass the tree, the light doesn't move at the speed of light plus 100 mph; it still goes at the speed of light. When you are moving—and this is the weird bit—time slows down and distances contract.

Suppose you measure a truck that is standing still and find it to be 16.4 feet long. If that truck now drives by you at 100 miles per hour, and you measure its length, you will find it has shrunk to 16.3999999999998 feet. Obviously, it would be absurd to worry about such small differences in day-to-day instances. But imagine the truck drove by you at 80 percent of the speed of light. Now it would be 9.84 feet long—a considerable difference. This may seem strange, but it gets a little bit stranger. While you (standing at a fixed point) measure the truck to be 9.84 feet long (as it is traveling 80 percent of the speed of light past you), your friend driving the truck performs the same measurement and finds that the truck is 16.4 feet. But this truck is a physical object, so shouldn't it have a definite length? Shouldn't physics be consistent?

The trick is that in order to measure the length of something, you have to do two things. You have to mark where the front of the vehicle is and then at some time later, you have to mark where the back of the truck is. When it is sitting still, this is very easy. You just pull out your measuring tape and walk from the front to the back. When the truck is moving at 80 percent of the speed of light, the fact that the two events take place at different times is important. Because time flows differently for your friend inside the truck compared to your stationary self outside of the truck. Even if you both mark the position of the front of the truck simultaneously, the moment each of you mark the location of the back of the truck will be at different times. This is what leads to the disagreement on the measurement of the truck. For those of you wondering, the "actual" length of the truck is 16.4 feet. (In physics, we call the stationary length measurement the "proper length.")

THE FORCE

THE FORCE

WHEN All films	
WHERE The galaxy	
CHARACTERS All	
PHYSICS CONCEPTS Force	

SHORT INTRODUCTION/BACKGROUND

The Force is central to everything in the Star Wars universe, yet it is mysterious. Characters with a high midi-chlorian (sentient microscopic organisms) count are described as being Force-sensitive. In rare occurrences, the Force itself can apparently conceive a child. How could an intelligent microscopic life-form allow a person to have magical brain powers? Can microscopic organisms have macroscopic effects on creatures? Is there actually a field that surrounds and penetrates all beings? Is the Force a specific incarnation of one of the four fundamental forces of the universe? Can it manifest as any one of the four? Or is it a fifth force entirely?

ELECTROMAGNETIC FORCE

Most interactions that we see on a day-to-day basis are electromagnetic interactions. Forces come in one of four varieties (as far as we know): gravity, electromagnetic, strong nuclear, and weak nuclear. The two nuclear forces only act on subatomic scales, so we do not experience them firsthand. Gravity is mostly only noticeable on very large scales. The force that your fingers exert on the book you are holding is created by the electrons in your fingers repelling the electrons in the book. The fact that you are not falling through the floor right now is because the electrons in the floor are repelling the electrons in your feet.

BACKSTORY

The Force has been described on several occasions throughout the films. The first comes from Obi-Wan Kenobi in *A New Hope* when he says, "It's an energy field created by all living things. It surrounds us and penetrates us; it binds the galaxy together." Later, we learn that there are microscopic organisms, known as midi-chlorians, which act as a channel for the Force to speak to Jedi. In *The Phantom Menace*, Qui-Gon Jinn says to Anakin, "Without the midi-chlorians, life could not exist, and we would have no knowledge of the Force. They continually speak to us, telling us the will of the Force."

THE PHYSICS OF STAR WARS

The Force manifests itself in many ways throughout the films. Most of these occurrences make it appear that the Force is an electromagnetic force. For instance, your lightsaber doesn't know whether you used the Force to bring it to you or somebody came along, picked it up, and handed it to you. The lightsaber feels a force that causes it to accelerate to its destination. That is the way an electromagnetic force behaves.

In physics we say that two things can only interact via a force if they both interact with that specific type of field. For instance, only objects with mass can interact gravitationally. Only charged objects can interact electromagnetically. Since the Force permeates everything, it is likely that everything in the Star Wars universe can be affected by Force powers. The description that "the Force" surrounds everything could be another way to say that there is a field with which everything interacts.

In theory, it would be possible for something to not interact with the Force and therefore be immune to its effects. For example, perhaps a Toydarian brain does not interact with the Force, thus rendering Jedi mind tricks ineffective on them. In our universe, dark matter does not interact electromagnetically. Since it does not interact with electromagnetism, it cannot produce or reflect any light (an electromagnetic wave), thus rendering it "dark."

> ### ENERGY AND GRAVITY
>
> Technically, two objects only have to have energy to interact grav-
> itationally. This is due to Einstein's famous energy-mass equiva-
> lency, $E = mc^2$.

Sensitivity to the Force could also allow a person to sense events in the distant past or in the future. Most laws in physics do not have a preferred direction of time, but as far as we know, there is no way to sense the future. There are theories that involve particles called tachyons, which can travel backward in time, but these have never been experimentally detected. Some descriptions of antimatter indicate that antimatter is just normal matter traveling backward in time. Perhaps the Force allows certain particles to travel through additional dimensions that connect different points in time (see Hyperspace section for additional information on extra dimensions).

THE PHYSICS OF REAL LIFE

As many of you are probably aware, there is no scientifically verified examples of magical brain powers. That's not from a lack of trying, though. Many people throughout history have claimed the ability to manipulate matter with their mind as well as be able to make predictions of the future through extrasensory perception. Perhaps one of the biggest names in this field of research is James Randi. He is a retired magician who had established the One Million Dollar Paranormal Challenge. In this challenge he promised $1 million to anybody who could pass a rigorous scientific test demonstrating paranormal abilities. It was terminated in 2015 after about fifty years of people failing to meet the scientific requirements of proof.

One famous test involving Randi was known as Project Alpha. In 1979 James McDonnell (of McDonnell Douglas) set up a parapsychology lab headed up by physicist Peter Phillips. They decided to examine the phenomenon of spoon-bending, the claim that people could bend spoons with their mind. James Randi was critical of the methodology used in the experiment and went as far as to hire people to stay within the protocols of the experiment, but to bend spoons (not with their mind) in a way that would generate a false positive. At the conclusion of the experiment, Phillips and his colleagues concluded that paranormal behavior could not be ruled out—at which point Randi revealed that he had gamed the system, thus discrediting the study.

Let's examine those midi-chlorians. They are described as microscopic organisms that live in every cell and can communicate to Jedi the will of the Force. How possible is that? Well, more possible than you may think. Studies into the microbiome of humans (the collection of bacteria living symbiotically with us) has been growing in recent years. Estimates of the ratio between the number of bacteria cells in a human body and the number of human cells are astonishing. The numbers range between 1:1 and 100:1 (meaning 100 bacteria cells for every human cell). A fact like this raises the question: what is a human organism? Is a human really a collection of many organisms in one? We have already discovered that our digestive system, in order to work effectively, relies heavily on gut bacteria.

FORWARD OR BACKWARD?

According to the current laws of physics, it is impossible to distinguish between normal matter going forward in time from antimatter going backward in time in a mirror. Next time you look at yourself in a mirror, imagine that your image is made entirely of antimatter going backward in time (it's not). Physics would be incapable of determining the difference between actual you and bizzaro you.

That's just with our stomach and guts; it couldn't affect our brain, right? Well, there are parasites that can take over animals (and possibly have effects on humans). A dramatic (and somewhat complex) example of this is the lancet liver fluke. It is a worm that lays its eggs in the intestines of cows. The eggs are excreted by the cows only to be consumed by snails. The eggs hatch in the snails, which then create cysts protecting themselves from the parasites and cough them up. The parasite balls are then consumed by ants. The parasites infiltrate the ants' brains and convince the ants that they should climb to the top of blades of grass near grazing animals (such as cows), thus completing the life cycle of the worms (and completing the lives of the ants). If you were to talk to an ant at this time, it would tell you that it had a biological imperative to climb to the top of that blade of grass. Perhaps the ant calmed its mind and listened to the microorganisms talking inside of it. Although there are theories that similar parasites have effects on humans, more studies are needed before conclusions can be drawn.

FORCE LIGHTNING

WHEN Episode VI, battle with Emperor Palpatine

WHERE Second Death Star

CHARACTERS Emperor Palpatine, Darth Vader, Luke Skywalker

PHYSICS CONCEPTS Electricity

SHORT INTRODUCTION/BACKGROUND

Lightning is a very powerful force of nature. It has dazzled humans for centuries. When Nikola Tesla created the Tesla coil at the end of the nineteenth century, he demonstrated the ability of humans to produce lightning and potentially harness its power. Like all scientific advancements, with great power comes great responsibility. Tesla coils are used for a few technical applications but are primarily used for entertainment and education. If used inappropriately, though, a Tesla coil could be lethal. Would it be possible to channel lightning through a human in the way that a Sith does it? What are the consequences of being struck by lightning? What causes lightning in the first place? It is important to note that throughout this section, potentially lethal behaviors will be discussed. Do not try any of this at home.

BACKSTORY

In the climactic battle between Luke and Darth Vader, Emperor Palpatine steps in to finish the job. Vader watches as the emperor repeatedly electrocutes Luke with lightning before stepping in and throwing his master into a reactor shaft of the Death Star. In this moment of redemption, we are able to see the power of Force lightning to both cause harm and make visible the skeleton of the person being hit. We also see the negative effects (sagging of the skin, teeth and fingers turning yellow, etc.) that Force lightning has on then-chancellor Palpatine as he fights Mace Windu with Force lightning. This Sith power demonstrates a powerfully destructive weapon.

THE PHYSICS OF STAR WARS

When the emperor first electrocutes Luke, he does it from about ten feet away. Assuming that the gas in the throne room is similar to that of air on Earth, the voltage needed to generate such an arc would be about 9 million volts (or about 6 million AA batteries, enough to fill about two freight containers). This is an immense amount of voltage to be generated, more even than US power lines. It should be clear that this voltage is too high for a normal person to channel.

One potentially problematic aspect of the depiction of the Force lightning in this scene is that lightning is both "lazy" and always in need of a place to go. The room appears to be made out of metal, which, as a great conductor, would likely attract the lightning to it before it could go all the way to Luke. If we assume that the floor is somehow not a conductor, then the Force lightning depicted here would be (essentially) a beam of electrons. Those electrons, unable to flow to the floor, would build up in Luke's body, possibly causing damage to his DNA.

In the battle between Mace Windu and Emperor Palpatine, Windu is able to deflect the Force lightning back onto Palpatine, causing severe disfiguration. The disfiguration mostly takes the form of wrinkles (as if he just aged a few hundred years). This is not the typical consequence of being struck by lightning. When lightning strikes a person, it can cause internal and external burns and possibly death. The primary cause of lightning-related death is an electrical current flowing across the heart. The heart is directed by a small electrical impulse (generated by the sinoatrial node). When large currents flow across the heart, this can disrupt (or destroy) the sinoatrial node, leading to death in most cases. Damage to the skin typically takes the form of burns or scarring, not wrinkles.

The lightning is always depicted emerging from the fingertips of the Sith. This is probably accurate. When it comes to lightning, it mostly connects with sharp, pointy things rather than flat things. This is because charge in a conductor will try to space itself out as much as possible. You might think, "Isn't a narrow finger going to have a bunch of charge bunched up?" You'd be partially right. Yes, that charge will be confined to the finger, but it will be able to move away from the large amounts of charge in the rest of the body.

When Darth Vader picks up the emperor, we see flashes of Darth Vader's skeleton through his suit. This may appear to be

---

LICHTENBERG SCAR

One type of scarring caused by large electrical discharges is known as a Lichtenberg scar. This looks like a scar in the shape of branching lightning. It is also possible to create Lichtenberg figures in glass and other materials, which look like lightning frozen in place.

---

cinematic exaggeration (and it is), but there is a way in which this would be possible. High-energy electrons (like those in a lightning bolt) that strike a metal target (like Darth Vader's suit) can give off X-rays. X-rays are used to image the skeleton. Before you get your hopes up about this, the only reason the results of radiography are visible is that those X-rays shoot through the body and on the other side there is film sensitive to X-rays. Some of the film receives exposure and some doesn't. Since calcium is a good absorber of X-rays, bones don't let X-rays through. Perhaps Darth Vader's suit is made out of photographic film? I suspect not.

Lastly, the electricity created by the emperor while electrocuting Luke appears to be a very low frequency AC current. When a person is electrified, the current flows through the muscles, causing them to contract. When someone is electrocuted with direct current (DC), the muscles tense up and cannot relax. When a person is electrocuted with alternating current (AC), the victim spasms at about the same frequency as the current. As Luke is electrocuted, he spasms back and forth at about a rate of once per second (or 1 Hz). Typical AC currents in US households alternate at 60 Hz, so the Force lightning is comparatively low-frequency current.

THE PHYSICS OF REAL LIFE
Despite being exceptionally common, lightning and the mechanism by which it works are not fully understood. High-speed footage of lightning (if you have never seen it, put this book down and Google it now) shows that immediately before a lightning strike, "leaders" are sent out in many directions. These "leaders" are sections of charged air particles that move outward, looking for an easy path to the ground (or other such place to discharge). When a leader from the cloud meets a leader emerging from the ground (or another cloud, etc.), a connection is made and large amounts of electricity flow through the path connecting the initial source to the ground.

The lightning bolt that we see is caused by the heating of the air to exceptionally high temperatures. During a lightning strike, the air close to the lightning is heated to temperatures around 36,000°F (or about three times hotter than the sun). At that temperature, the air turns to plasma and radiates a whitish-blue color (hence the color of lightning).

During a lightning strike, hundreds of thousands of volts are discharged through a bolt of current, which is around 20,000 amps. These numbers probably don't carry a lot of meaning to the average person other than "They sound big." Let us consider the different effects on a human body caused by approximate levels of current flowing through the body.

CURRENT (mA)	EFFECT
1	Slight tingle.
10	Painful shock.
100	Death. This is enough current to disrupt the electrical signals in the sinoatrial node. It is not enough to stop your heart, but enough to disrupt it enough to send you into fibrillation. Without medical assistance, this will cause death.
1,000	This is the current delivered by a defibrillator. This will cause your heart to stop, but your heart will restart itself and (probably) be okay.
2,000	Death. This fries the sinoatrial node and your heart cannot beat any more.
10,000	If a finger (for instance) experienced this much current, it would disintegrate before the current could make it through the rest of the body. Although this would lead to the loss of a finger (and all the hazards associated with that), it would not instantly cause death.

Given these facts, shouldn't a person be instantly vaporized by lightning? Well, no. The reports of 20,000 amps in lightning is assuming that it is traveling through air. If that same amount of voltage is applied to the body, less current will be produced because of the higher resistance of the body. Obviously, lightning can be lethal, but there are people who survive lightning strikes, some of whom seem to be mostly unharmed.

One possible cause for this survival is what is known as the skin effect. This is a tendency for high-frequency current to mostly concentrate itself on the outermost layer of conductors. For humans, the outermost layer is the skin, and thus it is possible for lightning to only go through the skin of a person before reaching the ground. Although very painful, this dodges the vital organs, which when damaged is what causes death.

FORCE JUMP

"Impressive...most impressive."
—Darth Vader (Episode V)

WHEN Episode V, Luke and Vader duel	
WHERE Cloud City	
CHARACTERS Luke Skywalker, Darth Vader	
PHYSICS CONCEPTS Force, conservation of energy, tensile strength	

SHORT INTRODUCTION/BACKGROUND

Force-users have many special abilities including Force lightning, mind tricks, and telekinesis. But what if a Jedi used telekinesis on herself? It has been demonstrated on numerous occasions that Jedi can jump to great heights. How could a Jedi use the Force to make this happen? How much would a Jedi's jumping ability be augmented by the Force? Is there a limit to how high a Jedi can jump?

BACKSTORY

In the famous duel between Luke Skywalker and Darth Vader on Cloud City, there is a moment when Darth Vader knocks Luke into a carbonite freezing chamber. Just before Vader can activate the mechanism, Luke jumps fifteen feet (according to the script) straight up and out of the chamber. This is the first demonstration of a Jedi using the ability now dubbed the "Force jump." The duel with the highest volume of Force jumps is "Duel of the Fates" in Episode I (between Darth Maul, Qui-Gon Jinn, and Obi-Wan Kenobi in the Theed Power Generator). Each of the duelers uses Force jumps to move from catwalk to catwalk; Obi-Wan uses a Force jump of sorts to ultimately defeat Darth Maul.

THE PHYSICS OF STAR WARS

Let us first look at the case of Luke leaping out of the carbonite freezing chamber to the relative safety of the carbonite hoses. In order to leap fifteen feet in normal Earth gravity, a person would need to leave the ground at about twenty-one miles per hour (a little bit slower than Usain Bolt runs). For Luke's muscles to accelerate him to this velocity in the span of, say, a tenth of a second, it would require a force of about 1,500 pounds (or a bit more than the force of the average soccer kick). Spread out over the area of his feet, this would cause a pressure of about 8.7 per square inch (the pressure at twenty feet underwater). Bones can survive pressures up to 22,000 psi (about twice the pressure at the bottom of the Mariana Trench), so a Jedi's bones should be able to survive this without issue.

If we work under the assumption that midi-chlorians don't enhance the tensile strength of bone, then we can estimate a maximum height to which Jedi can jump. Using Earth's gravity and rough estimates for Luke's mass and the cross-sectional area of his feet, we find that Luke could jump about 17,400 miles (or about five times the radius of Earth). This clearly indicates that the breaking strength of bone is not the limiting factor of how high a Jedi can jump (and anyway, who says the Force jump puts any of the force on the bones themselves?). So, what is the limiting factor? Most likely g-forces.

We often measure large accelerations as multiples of the acceleration due to gravity on Earth, sometimes referred to as one-g or g-force (the number familiar to all intro physics students, 32 feet/s^2). The average human will probably not pass out under a sustained force of 5g (this is what you experience on the most extreme roller coasters). Trained fighter pilots can sustain close to 9g acceleration. If we are generous and say that a Jedi can sustain a 10g acceleration, she could jump about sixteen feet (keeping the rough estimate that it takes a tenth of a second from the initiation of the jump to leaving the ground). If we give her half a second, the Jedi could jump a much more impressive four hundred feet.

The effects on the body from too many g-forces are (in order): gray-out, tunnel vision, blackout, G-LOC (loss of consciousness), and death. The hilarious implication of this is that a Jedi who hasn't had his full training may try to Force jump a little too emphatically and black out.

THE PHYSICS OF REAL LIFE

The most athletically fit humans can leap sixty-five inches upward. This means they could, from a standstill, jump up and land flat-footed on the head of a five-foot-five person. If we take this to be the current limit of human vertical jumping, what g-forces do those jumpers feel? They accelerate over a span of about half a second to reach a speed of about 18 ft/s, which corresponds to about a 1.2g force. It is unlikely that these athletes have any trouble with blacking out from jumping too high.

How can trained fighter pilots sustain up to 9g forces? Does the US Air Force specifically recruit people with a genetic disposition to being able to survive g-forces? Loosely speaking, blackouts occur when blood leaves the head and the brain is deoxygenated. To combat this, the air force has developed special suits with inflatable legs that can apply compression to the legs of the pilot to help hold the blood higher in the body. Also, pilots are trained to clench all of their muscles and breathe in very shallow, controlled breaths. If you want to waste an afternoon, you can watch videos on *YouTube* of pilots going through g-testing.

How do we know these things? Did scientists strap people into chairs, attach those chairs to rockets, and then see what happens? Well, yes. From the forties through the sixties, engineers were designing faster and faster planes that needed ejector seats in case of emergency. Most designs involved rocketing the pilot vertically out of the plane (which is already going at high speeds). This, along with the desire to launch people into space, got scientists thinking about how much is too much for a human to sustain. To test the limits of human acceleration, Colonel John Stapp was strapped into a chair on a long, straight track. On the back of the chair was a rocket engine. This engine accelerated the chair down the track and then was brought to a screeching halt. All of these tests were done so that we would know that it would be possible to send humans to space.

At this time, humans can't perform telekinesis. But if we did, could we exert those forces on ourselves? You may say, "Sure! Why not? It's a magic brain power, so anything is possible!" but let's consider a slightly related ability. You can use your arms to lift this book and presumably you can lift heavier objects. Can you pick yourself up? The problem with trying to do that is that the force your arms can apply on an object relies on your feet interacting with the floor. For our bodies to exert forces on external objects, there must be an equal and opposite force reaction. So, if the brain were able to exert a force on a block across the room, by the laws of physics, the block would exert an equal and opposite force on the brain. By extension using telekinesis to lift yourself would not be possible by the laws of physics because it would require pushing off from another object.

THE FROZEN BLASTER BOLT

"I'll show you the dark side."
—Kylo Ren (Episode VII)

WHEN Episode VII, opening scenes

WHERE Jakku

CHARACTERS Kylo Ren, Poe Dameron

PHYSICS CONCEPTS Newton's first law, phase changes, plasma, light

---

SHORT INTRODUCTION/BACKGROUND

In the Star Wars universe, the blaster is the all-purpose weapon of choice. As Han Solo puts it, "Hokey religions and ancient weapons are no match for a good blaster at your side." Blasters look like guns you or I might be familiar with, but shoot glowing beams (conveniently color coded to indicate your allegiances!). There is some debate as to whether the beams fired by blasters are made of gas, plasma, or maybe just light. In any case, when a blaster shot is frozen in midair, there is an interesting physics problem to explore.

BACKSTORY

In the seventh installment of the Star Wars saga, Kylo Ren introduces a possible new application of the Force when he freezes a blaster bolt in space (specifically, a bolt fired at him by the soon-to-be-captured rebel pilot Poe Dameron). We know that Force-users can alter the movement of solid objects (for example, by levitating them), but do these capabilities apply to any substance a Sith or Jedi might encounter? Based off of Kylo's suspended bolt and his grandfather's deflected blaster shots before that (shots coming from Kylo's own father), it would appear so. How might this be possible?

THE PHYSICS OF STAR WARS

Anyone who has taken a physics class should remember that an object in motion tends to stay in motion unless acted upon by an external force. If there's one thing that we'd expect the Force to be able to do, it's to exert an external force on an object. So, changing the motion of an object with the Force isn't anything new, but is a blaster bolt an "object"?

What are blaster bolts anyway? Various sources claim they are laser beams or some sort of plasma. Plasma is a state of matter (like solid, liquid, or gas) that is reached when you heat a gas to such a high temperature that the electrons escape the forces of their nuclei and roam freely. So, plasma is a highly charged soup of whatever substance you heated.

It is generally accepted that plasma blasters in Star Wars are powered by tibanna gas. Since this doesn't exist in the real world, we can't just look up its properties, but we can calculate some properties using basic science (and assuming they apply to fictional gases). If the bolt is made of very charged tibanna gas, Kylo would

CARBONITE, TIBANNA GAS, AND HAN SOLO

What do tibanna gas and Han Solo have in common? They are both safest to transport while frozen in carbonite. Tibanna gas is a fictional element found in gas giants around the galaxy, but most famously on Bespin. The gas is very volatile, which makes it useful for weapons and hyperdrives; in a pinch, you can spray your father in the face with it as a diversion. Due to high demand, it is also very valuable. One can learn more about the properties of tibanna gas by learning more about the planet where it is mined, Bespin. Given that Bespin is 118,000 kilometers in diameter and has surface gravity that is one and a half times that of Earth, one can estimate the density of Bespin as about 893kg/m^3. For comparison, this is about the same density as diesel fuel or two-thirds that of Jupiter.

THE FORCE TO STOP A BLASTER BOLT

In order to estimate the force required to stop a blaster bolt, we have to make a few assumptions and use our knowledge of plasmas. For most elements, the density of the plasma phase is not significantly different from the density of the gas phase. Since a blaster compresses the gas while heating it to the plasma phase, we assume that the plasma in the blaster has been compressed to a density approximately equivalent to Bespin's density. The blaster bolt itself can be measured to be about two-thirds of Poe's height and have the diameter of the barrel of the blaster. This means the mass of the bolt is around 1.89 kilograms or the weight of a textbook. Newton's second law tells us that a force is the change in momentum divided by the time it takes to stop the object. It takes Kylo Ren about half a second to stop the blaster, and in that half a second, it travels a bit over five meters. Plugging in the numbers, we can see it takes about 8 pounds of force to stop the blaster bolt.

only need to exert a modest 38 Newton force—the same as lifting an eight- or nine-pound object.

What if the blaster shoots lasers rather than plasma? Unlike plasma, light has no mass, and (as with all massless things) moves at 671 million miles per hour. We can slow down light by sending it through different materials; how much its speed changes is measured by a quantity called the index of refraction, represented by the letter n. For example, in glass (n = 1.5), light moves at about 450 million miles per hour, but in diamond (n = 2.4), it slows to about 280 million mph. So, if a Force-user is capable of changing the index of refraction of a substance significantly, the Force could possibly slow a beam of light traveling through that substance.

Either of these possibilities should mean big things for Force-users. Starkiller Base shoots projectiles that appear to be bolts of plasma, just on a much, much larger scale. If "size matters not" as master Yoda says, Kylo should be able to stop these projectiles just as easily. In contrast, the Death Star is a laser weapon, so a Force-user would have to change the index of refraction of a material to redirect or slow its beam.

THE PHYSICS OF REAL LIFE

Obviously, tibanna gas is fictional, so we're not going to experiment with it at any time, but we can do this with other materials. To scientists, manipulation of the motion of plasmas is commonplace. Perhaps the most famous example of this is at the Large Hadron Collider at CERN. In the LHC, materials are accelerated to amazing speeds (within a fraction of a percent of the speed of light) and collided together to make what is known as a quark-gluon plasma. In a typical plasma, the electrons and nuclei float around without sticking together. In a quark-gluon plasma, the matter itself melts into its most fundamental pieces.

The typical design of such a plasma containment chamber is a ring of magnets, which cause the plasma to flow in a circle. One specific example of such a device is a tokamak. Despite looking like the halls of the Death Star, a tokamak is a device designed to contain a plasma in a toroidal shape (think doughnut) for the purposes of generating energy from the process of nuclear fusion.

The joke about fusion power plants is that economically feasible fusion energy is always fifty years away, but recent developments in fusion technology have made it seem more promising. The MIT Plasma Science and Fusion Center announced in late 2015 that their modular tokamak design may be just what the field needs. Currently, the smallest tokamak devices are still large enough that a human can walk around inside. As such, it would be very hard to make a handheld blaster with current technology, but the size of fusion reactors is becoming smaller every year, so maybe we will finally have blaster technology in fifty years.

What about the controlling light? It turns out, we are exceptionally good at it, and only getting better. As far back as the

Assyrians in the seventh century B.C., we were making lenses to redirect light. Over the years, we have found new ways to manipulate light. Research groups have been able to slow down light to speeds around fifty feet per second, and in 2013 a group at the University of Darmstadt were able to stop light for about a minute.

There are other implications about the possibility of a Jedi changing the index of refraction of the air. For instance, when you see a mirage of a puddle on a hot roadway, it is actually light from the atmosphere that is taking a slightly bent path to your eye. The hot roadway is able to heat the air directly above it creating a small temperature gradient. Light prefers to move through hotter air (because of its lower index of refraction), thus light from the atmosphere will look as if it has come from the road itself! This means that if a Jedi could manipulate the air between him and the enemy, he could create a mirage of his body at another location. Similarly, a strong enough gradient could actually redirect a laser beam in another direction, causing all laser shots to bend around the intended target.

JEDI MIND TRICKS

"These aren't the droids you're looking for."
—Obi-Wan Kenobi (Episode IV)

WHEN Episode IV, Mos Eisley scene	
WHERE On the streets outside Mos Eisley cantina	
CHARACTERS Obi-Wan Kenobi, stormtrooper	
PHYSICS CONCEPTS Electromagnetism	

SHORT INTRODUCTION/BACKGROUND

The Jedi mind trick is the Force power that I wanted as a kid. As an adult, though, I recognize that ability to control people's thoughts is a terrifying prospect. It's a power that could lead to great improvements in the world or horrific destruction. It isn't possible for us to control a person's thoughts, is it? If we wanted to be able to control someone's thoughts, how would we go about that? Is it any easier to read a person's thoughts? Would it make sense that there are some people whose minds are too powerful for a Jedi mind trick?

BACKSTORY

On the dusty streets of Mos Eisley, a few stormtroopers stop Obi-Wan and Luke to inspect their droids. Obi-Wan, calm and cool as ever, explains to the stormtroopers that they have nothing to worry about and to move along. The stormtrooper complies and orders his men to move along. The Jedi mind trick is attempted a few other times with mixed success and consequences. Qui-Gon is unable to use the Jedi mind trick on Watto, because apparently Toydarians cannot be affected by them. Later, Obi-Wan convinces Elan Sleazebaggano to give up death sticks, go home, and rethink his life. In the grand scheme of things, it would have been way more convenient if Elan Sleazebaggano had been the one immune to the Jedi mind trick and Watto had gone home to rethink his decision to not sell the parts needed to repair the hyperdrive core.

THE PHYSICS OF STAR WARS

We don't know much about how the mind works and how thoughts are processed in the brain. What we do know is that when the brain is active, neurons allow ions to move in and out. This motion of charged particles will change the voltage (essentially a measure of how much charge is on either side of a membrane in this context) across the membrane of the neuron. When one neuron fires (allows the ions to flow), it might trigger an avalanche of neurons firing. This means that if we could control which neurons fired inside a person's head, we could potentially control his thoughts.

According to our current understanding of neuroscience, this is in principle possible. The brain is composed of about one hundred billion neurons, each connected to about ten thousand other neurons. According to current theories of neuroscience, some pattern of neurons fires, and based on the pattern, our brain is able to determine what we are seeing.

Dealing with numbers in the billions is going to be prohibitively difficult for this example, so we will pretend for a second that there are only one hundred neurons, and they are each numbered (1–100). In this simplified model, we could say that the response in the brain when a person sees the droids he is looking for is characterized by neurons 1, 37, 55, and 98 firing. When the person does not see the droids he is looking for, neurons 2, 75, and 79 fire. This means that if a Jedi could prevent neurons 1, 37, 55, and 98 from firing while also causing 2, 75, and 79 to fire, then a Jedi could in principle control the person's thoughts.

The Force can be used to move physical objects. Luke stacks rocks and lifts R2-D2. Yoda retrieves Luke's X-wing from the Dagobah swamp. Kylo Ren is able to freeze a blaster bolt in space. In order to accomplish these feats, an electromagnetic force needs to be exerted on the objects being moved. Electromagnetic force could also be used to cause ions to move across the membrane of a neuron. Admittedly, lifting an X-wing is significantly less difficult than the precision needed to move particular ions in a brain, but in principle this could happen.

So how could Watto be immune to such trickery? There are two possibilities. It is stated that this kind of trick only works on the weak-minded. So maybe the trick can only implant a thought in the brain and not replace other thoughts. A weak-minded person would just go with whatever comes into his brain and ignore the context immediately preceding it. A strong-willed person, though, would be suspicious about such a dramatic change in his thoughts. Another possibility is that Watto (and other Toydarians) has a brain that does not interact with the Force (see Force section). If his skull were a conductor, then electromagnetic forces would not be able to enter his brain.

THE PHYSICS OF REAL LIFE

Throughout history people have imagined what it would be like to have mind control abilities. Government organizations have even tried to study mind control to see if it is possible. Perhaps the most infamous example of this is the CIA's project MKUltra. It sounds like something out of dystopian fiction, but the CIA performed a number of studies on people to see if their minds were controllable. These tests included, but were not limited to, giving people LSD without their knowledge, hypnosis, sensory deprivation, and other forms of psychological torture. Outside of the short-term effects, there may have been darker consequences to these studies. Ted Kaczynski, also known as the Unabomber, was part of a study at Harvard (which he entered at age sixteen) in which participants were repeatedly verbally abused by a person (specifically using information provided about the participant's aspirations to belittle the participant). Although it is impossible to prove that this contributed to Kaczynski's mental instability, it certainly makes studies like these questionable.

On a lighter note, there are other ways to influence how people think. There is a branch of illusionist acts that use techniques collectively known as mentalism. The idea behind mentalism is that you can hypnotize or subtly influence people's decision-making. For instance, a mentalist might be able to convince you to pick box 1 rather than box 2 even though you feel like you have made a free choice. This is similar in nature to hypnosis, although hypnosis has received a lot more scientific study. For those of you worried that a hypnotist might convince you to do something horrific, scientific studies have shown that when you ask a person to stab somebody else (for instance), that the trance is broken and the subject won't do it. Hypnosis and mentalism are fields that have fallen out of favor since the days of Freud, but they do give us insight into the extent of the lack of understanding of how the brain works.

Researchers have been able to demonstrate that it is somewhat possible to read a person's mind. For instance, in a 2008 study researchers asked participants to hit a button with either their right hand or left hand while recording their brain activity. From looking at the recorded data, they were able to predict what the brain activity would look like immediately prior to the subjects' pressing the button with their left hand versus pressing it with their right hand. This means that when they asked participants to hit the button again, they were able to predict which hand would be used.

This may not sound very impressive. Surely our brain sends a signal to our hand, and there is a short delay before our hand actually moves, right? Well, in this study, the researchers demonstrated an ability to predict which hand was going to be used up to seven seconds prior to the person actually "choosing."

ROBOTICS

DROID MOBILITY

WHEN Episode IV, opening scene; Episode VII

WHERE Aboard *Tantive IV* above Tatooine; surface of Jakku

CHARACTERS C-3PO, R2-D2

PHYSICS CONCEPTS Rotations, magnetism, angular momentum

SHORT INTRODUCTION/BACKGROUND

For most people, moving from one place to another is second nature. Putting one foot in front of the other, changing direction, going up stairs, and stopping are all pretty much subconscious tasks. A robot, on the other hand, requires a lot of programming to be able to keep balanced and to direct its feet to stay upright. Different mechanisms have been devised to deal with this issue. What are some ways in which robotic balance is handled? How easy is it to create walking robots? Would R2-D2 and C-3PO be able to move as freely as depicted in the films? How can BB-8 move around while keeping its head on top?

BACKSTORY

As discussed in other sections, droids are everywhere in the Star Wars universe. Life as the characters know it wouldn't be possible without the droids assisting them. Although stationary droids can be helpful, a general-purpose droid is most helpful if it can follow you around. Different droid designs demonstrate a wide variety of mechanisms of mobility. C-3PO can walk on two feet. R2-D2 has wheels that allow it to roll just about wherever it needs to go (and rockets for when it can't roll). BB-8 has an entire body that can roll places. Other droids use some variation on these mechanisms.

THE PHYSICS OF STAR WARS

The mechanism of walking on two feet is the most difficult of all three of the mobility mechanisms demonstrated by the droids in

Star Wars. We don't really remember the months we spent as children trying to learn how to stand on two feet. For a droid, we need to program those calibrations directly in (or have a self-teaching AI). Our brains can, in a split second, adjust the relative pressure on the inside and outside of our feet, throw our arms out to the side, and tilt our whole body when we step on something uneven. C-3PO needed to have those types of reflexes programmed into him.

When walking through the hallways of *Tantive IV*, C-3PO would need a very stable, level surface to walk on. From a physics standpoint, staying balanced is a statics problem. If we consider C-3PO to function on a system where gravity is pulling down on his center of mass, there is an upward balancing force applied by the ground beneath each of the droid's feet. When there is an upward force under each foot (as there is while walking across a flat, smooth surface), balance is easy. When the ground is uneven, the force it applies will be in a direction other than straight up. When walking across a sand dune on Tatooine, for instance, the sand shifts underfoot in order to provide a mostly upward force. But if C-3PO were to walk on a steep rock face, this droid's flat-bottomed feet would be unable to bend like human ankles, and he would likely fall over.

R2-D2 has three "legs" (with mounted wheels) that provide primary movement. This makes it much easier for the droid to keep stable. Having three points of contact with the ground allows for one point to be removed from the ground while still maintaining balance without much adjustment. The steering of R2-D2 is just like steering a tricycle; the rear wheels always pointing forward while the front wheel can rotate. I hope that R2-D2 has all-wheel drive because otherwise it would be easy for this droid to get stuck in the Tatooine desert.

BB-8's ability to roll everywhere while maintaining its "head" on top the whole time is a feat of engineering. It would be easy to write this all off as CGI magic, but there are real models of

BB-8 that can do this. How could this work? The design team is mostly silent on this topic, but it can be done with gyroscopes and magnets. Imagine the underside of BB-8's head is a bunch of ball bearings, which are free to roll on the exterior of the body. It would be easy then for the head to move around the body (or have the body move freely underneath it). But if there is a magnet inside of the body that holds the head in place, the body can roll while the head remains in place. As long as that magnet is always pointing upward, BB-8 won't lose its head.

The trick arises when you consider how to keep BB-8's head balanced. There must be an interior structure of the body that keeps the magnet (which holds the head) always pointing away from the ground. Then the exterior of the body needs to be able to roll freely. The head also needs to be able to roll freely relative to the body, but be held in place by the magnet. This whole time, the magnet needs to not affect the ball bearings or the rest of the freely rotating body. This can be accomplished with a gyroscope (think of a top). When spinning, a gyroscope resists being reoriented. This means that if there is a spinning gyroscope inside of BB-8 with a magnet attached to its end, it would stay upright even as the rest of the body moved around it.

THE PHYSICS OF REAL LIFE

The tech team on *The Force Awakens* made several BB-8 models for different tasks that the droid needed to perform. So, we are more than capable of actually producing a robot with mobility such as that of BB-8.

The mechanism for controlling rotations demonstrated by BB-8 is used by NASA all the time to reposition satellites. Think about trying to point a satellite (imagine the Hubble telescope) at a target for observation (consider the Pillars of Creation). The Hubble is constantly orbiting Earth, and Earth is constantly orbiting the Sun while the Pillars of Creation are thousands of light-years away. This means that it is necessary to rotate the Hubble in order to maintain its orientation relative to the Pillars of Creation. If you have seen one image from the Hubble telescope, it is probably of the Pillars of Creation. They are clouds of interstellar dust a few thousand light-years from Earth. These clouds of dust will give rise to stars as the dust coalesces into spheres, hence the name.

The simplest way to accomplish this would be rockets. The problem with them is twofold. Rockets rely on fuel, so once the fuel runs out, the mission is over. Carrying fuel to space is also very expensive and limits the size of the satellite. The second issue with rockets is that using fuel creates a cloud surrounding the satellite which impedes the view of the Pillars of Creation. Because of all of these issues, we use gyroscopes (or in this context, they are referred to as reaction wheels).

When a system doesn't have external torque, its angular momentum must stay constant. For BB-8 to roll forward, something inside of it has to roll backward (or distribute its weight differently inside the body) to conserve angular momentum. For a satellite, it is necessary to have at least three reaction wheels (one for each axis of rotation). The computers on the satellite can then calculate how much to speed up or slow down the rotation of each wheel in order to generate the needed rotation to keep the Hubble pointed at the Pillars of Creation so that it can take a beautiful picture.

Bipedal robots are being designed as well. ASIMO is a robot designed by Honda that is bipedal. It has many other capabilities that are not that impressive for a human but are very impressive for a robot. ASIMO can walk forward and backward; it can walk up and down stairs; and it can even kick a ball. As it performs all of these tasks, it needs to constantly reposition its center of mass over its feet in order to maintain its balance. It has taken more than thirty years of development to bring ASIMO from a pair of legs that would fall over when walking to being the robot it is today.

Some may picture the development of robots as a frivolous luxury enjoyed by families like the Organas. There are very real and important applications to using robots today. For instance, during the Fukushima disaster in 2011, it was important to assess the status inside the reactor area. This area was covered in debris and was highly radioactive. Robots were used to perform some of the tasks, but they did not have the mobility to move through the debris and were not able to fully assess the situation.

AI

"I am C-3PO, human-cyborg relations."
—C-3PO (Episode IV)

WHEN Episode IV, opening scene	
WHERE Aboard an ambassador ship	
CHARACTERS C-3PO, R2-D2	
PHYSICS CONCEPTS First principles, quantum field theory	

SHORT INTRODUCTION/BACKGROUND

The Star Wars saga heavily relies on droids for plot advancement as well as comic relief. Nearly every droid featured in the movies appears to have a unique personality. R2-D2 seems bold and sarcastic, C-3PO appears bumbling and unaware, and BB-8 comes off as sassy but loyal. Regardless of the individual personality, the droids all seem to have some form of intelligence and emotional maturity. Can a droid experience emotion? Can circuitry mimic the complexity of the conscious mind? How can we be certain that a computer is truly conscious? For that matter, how do we know that we are conscious? Are we just simulations in a very advanced computer?

BACKSTORY

From the opening scene of *A New Hope*, droids are an important part of the Star Wars story. The first conversation between two characters onscreen is between R2-D2 and C-3PO. This highlights the importance and the ubiquity of droids in the Star Wars galaxy. They play integral roles, from halting a garbage compactor to transporting the magnetic tape with the Death Star plans. It appears that having a droid is central to life in the galaxy. All strata of society seem to have access to them from the simple moisture farmers on Tatooine who are able to afford many droids to the Death Star, which is crawling with them.

THE PHYSICS OF STAR WARS

It's hard to find a plot element that is not somehow influenced by droids. Droids have varying levels of intelligence and functions that correlate with the level of intelligence. If we rate intelligence solely on the amount of dictionary-style knowledge conveyed during the movies, C-3PO probably is the most "intelligent" droid we encounter. Knowing more than six million different forms of communication and being able to translate between them is an impressive feat. Since the intelligence displayed by C-3PO can be roughly described as "book smarts," his role is in diplomacy and conversation rather than in battles. Despite this, he seems to find himself in a number of dangerous situations over several decades. R2-D2 appears to have more practical knowledge. To put it in physicists' terms, where 3PO is the theorist, R2-D2 is the experimentalist. When there's a need to open a door or stop a garbage disposal, R2-D2 is able to get the job done.

A different style of intelligence is demonstrated by the droid armies of Episodes I and II. These droids are programmed to respond to orders and understand basic military techniques. They do not have the capabilities of free thought or improvisation in the way that C-3PO and R2-D2 appear to. The droid army is exceptionally good at standing in formation and fighting, but on several occasions, their lack of intelligence is demonstrated. For instance, when 3PO's head is attached to a battle droid's body, it continues to execute fundamental operations like marching and fighting, but seems not to realize that it is missing a head.

Further down the list is the droid factory, found on Geonosis, which constructs the droids. The factory has an automated assembly line with a single function: build battle droids. Perhaps if given other parts and a different programming, it could construct other droids, but it is very inflexible. It seems to have a single operation that it can repeat efficiently and repetitiously. Smaller variations (such as having the separated head and body of C-3PO) in its line does not throw it off, but if either of those pieces had been out of place or in a different orientation, the assembly line probably would fail.

One could also argue that the automated warning systems on the ships are a form of artificial intelligence. Sensors all over a ship measure things such as temperature, pressure, and current (to name a few) and report if something is out of order. If those sensors detect that there is a leak in the ship, they will communicate with a warning light or siren. Although this is a primitive form of intelligence, the argument could be made that the ship's computer is taking input from the surroundings, analyzing it, and then acting (or not) based off the inputs. How is this any different from what we do?

THE PHYSICS OF REAL LIFE

If you accept the lowest level of artificial intelligence described previously, then we have achieved AI. The thermostat in your home has that level of intelligence, but this isn't terribly exciting. Do we have computers that can imitate a human? If we did have such a computer, how could we be sure it was conscious and not just a really good faker?

Enter Alan Turing. He developed a test (which now bears his name) to determine if a computer can imitate a human. Roughly speaking, the test involves a person typing questions into a terminal and responses coming back on the terminal. Sometimes the responses are written by a human, sometimes they're provided by a computer. If the person at the terminal cannot reliably determine whether he or she is speaking to a computer or a human, then the computer is said to have passed the Turing test. As of the writing of this book, nothing has passed the Turing test, but algorithms are edging closer and closer to being able to do so.

Let's imagine for a moment that a computer passes the Turing test. Is the computer conscious? Is it alive? Does that computer have emotions? There may not be a definitive answer to these questions. From a physicist's standpoint, reproducibility of results is key. If a computer passes the Turing test just once (that is, it tricks one human), it might be a bit hasty to call it conscious. If it were able to pass the Turing test a hundred times, a thousand times, a billion times? It may have more credibility at that point, but that is no guarantee of consciousness.

If we assume that humans are conscious beings, what is it fundamentally that makes them conscious? A physicist might say that it has something to do with the way that ions interact in the brain while neurons communicate input and process information. When we think, the neurons in our brains will reach a threshold voltage above which they fire a signal to the next neuron. This firing may trigger an avalanche of neurons firing to create a thought (whatever that physically is). The interaction of ions can effectively be described by electromagnetic theory and follows a basic algorithm: if some number (membrane voltage) is greater than a certain level (threshold voltage), then open sodium-potassium channels. If this is the type of procedure underlying a thought, then how are our brains any different from that of a computer?

A computer has transistors that can either allow current to flow or not depending on whether the signal sent to it (most likely from another transistor) is a 0 or a 1. Does this mean that there is some more abstract, inherently nonphysical part of what makes us conscious? Most cultures have a name for this idea.

One proposed theory (which is essentially untestable, making it inherently nonscientific) is that there is a consciousness "field" in the way that there is an electromagnetic field. This would mean that just as an electron and proton are different excitations of the electromagnetic field, a human brain somehow triggers an excitation in the consciousness field, something a rock is not able to.

If we start with the assumption that it's possible to program a computer to have conscious thought, then it's just a matter of computing power before we can simulate conscious beings. Simulating billions of conscious beings would require a lot of development and creation time, but, if it is possible, then it's just a matter of computing power to simulate multiple universes. In this scenario, there would eventually be a "real" universe with billions of universe simulations inside of it. Of all of these possible universes (including the real one and the simulations), it is statistically more likely that you currently inhabit one of the billion simulations rather than the single "real" universe.

Does this have any real consequences for your day-to-day life? No, but it's fascinating to think about.

ROBOTIC COMMUNICATION

WHEN All films	
WHERE The galaxy	
CHARACTERS R2-D2, BB-8, C-3PO, K-2SO	
PHYSICS CONCEPTS Information theory	

SHORT INTRODUCTION/BACKGROUND

When programming a droid, it is important for the droid and the human user to be able to communicate effectively (or for the droid to communicate with other computers). Strictly speaking, giving a droid a personality is not a requirement, but it can make interacting with it more natural. How can we program a droid to communicate with a human or another droid? Is spoken language the most efficient method of communication? Do the droids in the films communicate in ways in which we'd expect real droids to communicate?

BACKSTORY

Droids are constantly communicating with one another and with non-droid characters throughout the Star Wars films. Reporting the odds of survival in dangerous situations and expressing distrust of new people are just a few services that droids provide to characters in the films. R2-D2 and BB-8 use their characteristic chirps and beeps, C-3PO has its six million forms of communication, and K-2SO only really speaks stoic cynicism. Depending on the droid's purpose, it will have a range of communicative abilities.

THE PHYSICS OF STAR WARS

All forms of communication have to contain information. Whether that is encoded into words, which are combinations of letters or symbols, or bytes, which are combinations of zeroes and ones, there has to be a way to characterize that information in a way that both parties can understand. If I say, "01101000 01101001," it is not immediately obvious that that is binary for "hi." To a computer, that binary code would be immediately obvious.

Humans usually grow up learning a spoken language or languages specific to their geographic region. It would make sense, then, to program droids to speak the language of a local area (or have a translator droid). It is possible that humans would just learn to speak their droid's language. It is clear that certain people who work with droids regularly also learn their language. On several occasions, for instance, Luke responds directly to R2-D2's beeps.

K-2SO seems to have the most fleshed-out personality of all the droids in the Star Wars universe. Speaking whatever comes into his circuits is apparently a by-product of his reprogramming. Although this is described as a flaw, it is very likely that droids would "speak their mind" without consideration of whether a human wants to hear it. Nuance is a very difficult thing to program. To a robot, providing complete information so that the best decisions can be made sounds like a useful service. Hearing that the odds of successfully navigating an asteroid field are 3,720 to 1 is not useful to a risk-taker like Han Solo. Han makes decisions with his gut rather than with raw data. This concept would be foreign to a droid.

If the droids in Star Wars are programmed at all like robots in our universe, they have to make decisions on quantifiable data. Since, at the heart of it, all coding is zeros and ones, all information stored in a computer has to be able to be described by a number. What color is C-3PO? R = 249, G = 239, B = 168. What are Luke's chances of survival in Hoth's cold climate? 725 to 1. What are the odds that Jyn Erso is going to use that blaster on you? It's high. It's very high.

R2-D2's ability to communicate with every computer system on every ship is a subtle point in the movies, but it speaks volumes for how technology works in the Empire. With a single appendage, R2-D2 can stop a droid factory from dumping molten metal on

Queen Amidala or stop a trash compactor on the Death Star. This may seem like a cheap plot device to allow characters to get out of sticky situations, but it actually says something about the standardized systems of the Empire. How many times have you thought to yourself, "Do I need a USB mini or USB micro?" or "I can't open that file as it's Mac-formatted, not PC." With the standardization of technology in the Empire, that isn't an issue!

THE PHYSICS OF REAL LIFE

Teaching language to computers is something that we have striven for since the dawn of the computer age, but we're just making it a reality now. There is an increased prevalence of voice-commanded devices such as the Amazon Echo, Microsoft's Cortana, Apple's Siri, or Google's Home. These devices record what is spoken to them, interpret the command in real time, and then act accordingly. Currently, they are able to follow simple commands like, "Play *The Empire Strikes Back*" or "Remind me to update Wookieepedia." If you were to try to have a philosophical discussion with one of those devices about the merits of going to the dark side, it would be at a loss. It could hear and parse the words, but it would be unable to assign meaning to them.

Perhaps if we could speak the language of computers directly, though, we could communicate better with our devices. We would

EXPERIMENTAL LANGUAGES

Some experimental languages have been developed by people such as the linguist and philosopher Noam Chomsky. He looked at hypothetical languages where there is only one character and each idea is represented by a different number of that character all the way to a language where every concept had a single, unique symbol. At this time, there isn't an agreed upon ideal balance between those two extremes.

HAMMING CODE

There is a concept in computer science called a Hamming code, which is designed to be able to catch errors as they happen. A perfect Hamming code uses every combination of ones and zeroes as either a message or an error vector that points back to the correct message. An example of a perfect Hamming code is a code where you want to convey either "yes" or "no." If "yes" is encoded as 111 and "no" is 000, then even if one bit is messed up, you can still know what the message is meant to be. Since nothing is truly perfect (despite the name), a perfect Hamming code can only correct a single-bit error.

have to be much more precise with what we say. Computers don't have time for idle chatter, and they certainly don't do well with errors in speech. When a human confuses two words while speaking to another human, context can often make it clear what is being said. But if you are speaking to a computer in binary, the difference between "01101000 01101001" and "01101000 01100001" is the difference between greeting the computer with "hi" or laughing at the computer with "ha." This may seem like a silly example, but being precise in communication is important in all contexts, but with computers it is especially important. Saying "The suspense is killing me" makes sense to a human, but a robot may interpret that as somebody attempting murder.

Google Translate is on the leading edge of teaching computers language. Google has started to use neural networks to evolve the translate functionality. This is not a biological system, but a computer system that is designed to mimic the brain. A set of data is provided to the computer along with the desired output for that data. The computer then begins to "learn" what factors are most important for determining meaning. From this methodology, Google Translate has become significantly more natural in its translations than its predecessors.

MACHINES MAKING MACHINES

WHEN Episode II

WHERE Geonosis droid foundries

CHARACTERS Padmé Amidala, Anakin Skywalker, C-3PO, R2-D2

PHYSICS CONCEPTS Automation, singularity, blackbody radiation

SHORT INTRODUCTION/BACKGROUND

Throughout history humans have been developing tools. Early in history this consisted of simple tools such as wheels and axles, and now we have tools in the form of robots that can build full cars. Eventually we might be able to develop robots that can build copies of themselves. The point in which machines can self-replicate is both a goal of automation and a potential turning point. Once machines can make machines, what will stop them from constructing a robot army to take over the world? With the advent of 3-D printers, machines self-replicating is becoming more and more possible. This concept is related to the singularity: a point at which artificial intelligence outpaces human intelligence.

BACKSTORY

While trying to save Obi-Wan Kenobi on Geonosis, Anakin, Padmé, C-3PO, and R2-D2 stumble upon the underground Geonosis droid foundries. They see many of the steps in creating a droid army, from smelting metals to soldering heads. This is an entirely automated factory, which produces an entire droid army with little input from sentient beings. How would a factory such as this operate? How much energy would it take to operate such a factory? Would there be enough raw materials to accomplish this? What happens if there is a glitch? Does the whole thing stop, or does it run quality control automatically as well?

THE PHYSICS OF STAR WARS

From what we can see in the film, the process begins with molten metal of some kind, which is loaded into a ladle via a piping system. From there, it undergoes processing and eventually emerges in small (still-warm) bricks. Those bricks are then stamped into the pieces of the droid, which are welded together.

Let us start with the metal. We can get a sense for its melting point since it is in solid form when glowing a reddish orange whereas in its liquid state it glows a bright yellow color. When a material is heated to the point that it glows reddish orange, its temperature is around 2,200°F whereas when it is bright yellow, it is closer to 6,000°F. This means that the melting point of that metal is somewhere in between. This is perfectly reasonable as most metals have a melting temperature in the range of a few thousand Fahrenheit (tungsten is the highest at 6,150°F).

How much energy would this process take if we were making the droids out of iron? It would require the burning of about two gallons of gasoline to heat up enough iron for each brick we see as Anakin fights the Geonosians. It stamps two of these bricks into a shape about every five seconds. Using this rate of production, just the heating of the metal part would require about 10 percent of the output of a typical coal plant. Factoring in all the power for the

---

BLACKBODY RADIATION

We can relate temperature to color because of a process known as blackbody radiation. The wavelength of light that is emitted most by a hot object is related to temperature via Wien's law. The gist of Wien's law is as temperature goes up, the most common wavelength of light decreases. This allows us to know the temperature of an object from its color.

soldering irons, the conveyor belts, and all the robotic pieces working on the droids would send this number skyrocketing.

It is also clear that the foundry does not have much quality control. At one point C-3PO's head is knocked into the conveyor belt of heads to be attached to droids. To the machine's credit, it is flexible enough to be able to pick up a head that does not fit the specifications of a B1 battle droid. A well-maintained factory, though, would surely recognize that something was wrong with that unit and dispose of it appropriately. Similarly, when C-3PO's body appears in the lineup, it should not be allowed to pass. Likewise, Anakin's hand (and a piece of a robot) is trapped underneath a sheet of metal, which is stamped down from above. Quality control should have immediately been called to that site to remove the faulty piece.

There are a number of concerning things about Anakin's arm being trapped by a sheet of metal. First of all, in order to bend a sheet of steel in that fashion, it would require at least a thousand pounds of force. Applying such force to a forearm would at least break his arm if not much worse. But even if this metal is significantly more malleable than steel, the rest of the assembly line seems to work around the large deformities in the sheet of metal. This is largely inconsistent with the basic principles of an assembly line. Uniformity of pieces is crucial. When one piece has a large aberration, it needs to be removed before it reaches the next stage. If the machines can sense the aberration, they should be able to remove the faulty piece, as it will not work for a droid.

THE PHYSICS OF REAL LIFE

There is a trend in manufacturing to move away from human labor and replace it with automation. This is exemplified well in the manufacturing of cars. In the early 1900s Henry Ford used the assembly line to revolutionize the production of automobiles. It is hard to find direct comparisons of numbers, but to give a sense: At its peak, the Highland Park Factory employed close to 48,000 people.

As of 2012, the Dearborn plant was down to about 3,200 people. This is all while a plant in the early 1900s could turn out about 1,500 cars a day whereas a plant today can produce almost 500,000 vehicles per year. Per person, this is about a tenfold increase in productivity.

Automation in manufacturing is also being assisted by 3-D printing of parts. A 3-D printer can take a computer model of a three-dimensional object and print it out. There are many different varieties of 3-D printers, but two common types are extrusion and powder bed. An extrusion 3-D printer lays down layers of a material to slowly build a model. A print head moves back and forth depositing molten strings of material, which quickly cool and solidify to create the model layer by layer. It looks a lot like a machine designed to ice a cake that has been given plastic instead of icing.

In powder bed 3-D printing a layer of powder is set out evenly to start. A nozzle (much like the one used in extrusion 3-D printing) will place a type of glue to hold certain grains together. Once the first layer is complete, another layer of powder is laid down. The adhesive is then applied where necessary to the next layer, and the process continues. At the end, excess powder can be removed from the model and the bed, and it can be reused for the next procedure.

Both styles of 3-D printing discussed here are additive. This means they add material to build a model rather than selectively remove material (much like a sculptor carving a statue). Subtractive 3-D printing does exist, but it has limitations in that it cannot produce hollow models without needing to patch a hole in the end. It is also worth noting that 3-D printing can use resins, polymers, metals, food, and many other types of materials to build pieces.

At this time, 3-D printers are used to make precisely designed parts in large numbers, which are then assembled elsewhere. It is not too big a leap to imagine a line of 3-D printers that produce all the parts needed to make a 3-D printer. Those parts are then sent on a conveyor belt to a series of robots, which can assemble those pieces into a working 3-D printer.

ROBOTIC HEALTHCARE AND PROSTHETICS

WHEN	Episode V, concluding scenes
WHERE	Aboard *Redemption*
CHARACTERS	Luke Skywalker, Darth Vader
PHYSICS CONCEPTS	Torque

SHORT INTRODUCTION/BACKGROUND

Advancements in medical science have prolonged life and made improvements to quality of life. Nevertheless, it appears that medical science in the Star Wars universe is more advanced than us in both life support systems and prosthetics. How realistic is the prosthetic hand that Luke receives at the end of *The Empire Strikes Back*? Could we design a Darth Vader–style suit to provide life support to a critically ill and injured patient?

BACKSTORY

Throughout the Star Wars saga, we get brief glimpses into how healthcare works. Luke nearly freezes to death on Hoth, but is resuscitated in a bacta tank with the help of medical droids. Later, Luke receives a prosthetic hand to replace the one severed by Darth Vader in their lightsaber duel. Darth Vader himself is described as "more machine than man." His suit is designed to provide both prosthetic limbs and life support to keep what is left of Anakin alive. The most extreme version of this is General Grievous, who has had most of his biological parts replaced with mechanical parts.

THE PHYSICS OF STAR WARS

When Luke has a prosthetic forearm and hand attached to what remains of his biological arm, the medical droid performs tests to ensure he has a full range of motion and feeling. In order to have control over the device as well as receive sensation from the prosthetic, the wires must interface with Luke's nervous system. In principle, something like this is not difficult to do. In practice, that is a different story.

The nervous system works by sending electrical signals from the brain to the muscles (for motion) or from the extremities to the brain (for sensation). These signals take the form of action potentials. In order to wire a prosthetic into the nervous system, the wires of the prosthetic need to be connected to the ends of neurons (known as axons). This way, when the neuron fires, it will transfer its electric potential to the wire causing current to flow and moving the extremities. The reverse process can be done for touch sensation.

CELLULAR ELECTRICITY

An electric potential is a separation between charged particles. In the case of a neuron, the primary ions at work are sodium and potassium (both positively charged). The cell membrane only allows potassium to cross over it with ease. The inside of the cell is mostly filled with negatively charged particles. In addition to this, for every two potassium ions that enter the cell, it will use pumps to push three sodium atoms out of the cell. This process maintains what is known as a resting potential.

When a neuron wants to send a signal to another neuron, a spike in electric potential across the membrane occurs (known as an action potential). This spike is caused by the cell opening all of its sodium channels, which allows sodium to flow freely from outside the cell to inside of the cell. Once the concentration of sodium on either side of the membrane balances, the potassium channels open, allowing the potassium to flow out of the cell. At this point the potassium channels close and the neuron begins pumping sodium back out of the cell to restore it to its resting potential.

As for Darth Vader's life-support armor, it has its own engineering problems beyond the challenges with prosthetics described previously. Although ventilators and intravenous nutrient/medicine delivery systems are commonplace in hospitals, they require input of raw materials. For instance, where does Vader's air come from? Presumably when on the Death Star or in a Star Destroyer, his ventilator could draw air from the surroundings. If in a harsher environment, the suit needs the capability of either recycling the gases that he exhaled or having a tank of gas constantly in tow. Considering that there does not appear to be such a tank (and that being repeatedly struck by lightning did not cause catastrophic combustion), the former suggestion is probably the case.

General Grievous is the most mechanical of all cyborgs in Star Wars. Grievous is not much more than a heart and brain with a robotic skeleton. He is just about at the limit of what can be called a Kaleesh with robotic enhancements. In order to prevent his organs from rupturing, they must be contained in a pressurized sack. Internal organs have counterpressure on them at all times from surrounding organs and the skeletal system. Without that counterpressure, they would erupt like a popcorn kernel. The question of whether Grievous is a robot with biological enhancements or a Kaleesh with robotic enhancements could leave philosophers arguing for days. Does the answer lie in the current percentage of an individual's makeup? Or does the trajectory of an individual matter more (started as Kaleesh and replaced biological parts with machine parts)? These are not questions with definitive scientific answers.

In order to accomplish these feats of medical science, precise and knowledgeable medical staff are needed. The medical staff in the Star Wars universe appears to be entirely droid-based. When Anakin's life is being saved (thus completing his transition into Darth Vader), we see droids that are applying artificial limbs, cybernetic implants, providing blood transfusions, and doing surgery. The same style of surgical droid assists Luke to recover after his time in the harsh cold of Hoth as well as in testing his prosthetic hand.

THE PHYSICS OF REAL LIFE

It may seem like a fantasy to have robot doctors, but strides are being made to accomplish just that. IBM developed a computer referred to as Watson that is capable of making medical recommendations. Watson is able to access medical studies, potential drug interactions, patient history data, etc. to recommend a treatment option. As with all medical recommendations, complete knowledge of a situation is impossible, so Watson's recommendations are given a confidence rating. For instance, this means that Watson will recommend something like: With 90 percent confidence, this lung cancer patient should undergo chemotherapy. With 5 percent confidence, this patient should undergo surgery. Although Watson is only available in limited situations at this time, perhaps someday healthcare will be completely provided by robotic diagnosticians. There are already surgeries in which a doctor controls robotic arms that have significantly better precision than humans. Someday, those robotic arms will be able to sense on their own where they need to operate.

The technology of prosthesis is not quite at the same level as seen in Star Wars. We're just discovering at this time how to interface prostheses with a user's brain. Wires connect the prosthesis to electrodes attached to muscles, which still work in the body. When the user thinks about contracting that muscle, the electrodes record the signal and start to learn the patterns of brain activity. A small computer in the prosthesis then learns to pay attention when the user contracts that muscle and maps the signal to some functionality (like touching the thumb and forefinger together). For example, a nerve that was previously used to move your fingers is redirected to attach to a muscle in your chest. The prosthetic will then sense when that particular chest muscle contracts and move the fingers of the prosthesis appropriately. This procedure is known as targeted muscle reinnervation.

Most people are probably familiar with basic life support systems that are currently in use in hospitals. There are ventilators, feeding tubes, IVs, etc. What is the next generation of life support? Well, there is current research into how to support life on long-duration space flights. One such project is known as Micro-Ecological Life Support System Alternative (MELiSSA). This initiative is part of the European Space Agency. This system, when operable, will allow for zero waste as everything (and I mean everything) will be recycled for use in some way. MELiSSA consists of four compartments. The first compartment liquefies everything. Proteins are broken into amino acids, cellulose is broken into smaller polysaccharides, and other waste is converted into simple molecules such as hydrogen, ammonium, and CO_2. The second compartment contains organisms that will break down the volatile fatty acids remaining after the first compartment using the CO_2 and light. The third compartment has organisms that can convert the ammonium into nitrates, preparing the nitrogen to be fed into plants via fertilizer. The final compartment uses algae to generate oxygen and food for the astronaut, thus completing the cycle of reuse of waste. If this could be miniaturized, this could be the perfect design framework for a Darth Vader suit.

OTHER TECH

GUIDANCE COMPUTERS

WHEN Episode 1, Battle of Naboo	
WHERE Naboo	
CHARACTERS Anakin Skywalker, R2-D2	
PHYSICS CONCEPTS Circular motion, gravity	

SHORT INTRODUCTION/BACKGROUND

Finding your way from one side of town to another can be tricky at times; finding your way across a galaxy is even harder. In order to chart paths, whether through hyperspace or when taking off from Naboo and going into low orbit, guidance computers are crucial. How do they work? What goes into calculating an interstellar journey? Is it any easier to plot a course to low-Naboo orbit than it is to plot a trip to Coruscant? What are current guidance computers capable of?

BACKSTORY

Guidance computers may not be necessary for dusting crops, but as Han Solo said, "Without precise calculations we could fly right through a star, or bounce too close to a supernova." Calculating the path through hyperspace requires precise work, and one mistake could be the end of you. Similar levels of precision go into taking off and landing. If you come in too fast or too steep, that would also end your trip in a moment's notice. Although the autopilot successfully helps Anakin take off in the Battle of Naboo, Anakin has R2-D2 shut off the autopilot before he can get shot down.

THE PHYSICS OF STAR WARS

When looking at a map of the galaxy, plotting a course from Tatooine to Alderaan or Coruscant to Naboo may look like an easy enough straight shot. Perhaps when going through hyperspace, it can be a lot more direct than through real space. Either way, it takes time to make it from one planet to another. In the time that you're traveling, you have to take into account where your destination is at this moment as well as where it will be when you arrive. Because everything in a galaxy is orbiting the center of the galaxy, your destination will move as you travel toward it. You may actually have to target a point in space that is empty at the moment, knowing that by the time you arrive, the planet will also be arriving. It's kind of like meeting a friend for coffee. You both come from other locations and meet at a third. Your friend, in this analogy, just happens to follow a predictable elliptical path.

If the Star Wars galaxy is anything like the Milky Way, Coruscant, being close to the center, is moving around the galactic center at a different speed than Tatooine (which is on the outer rim). Stars orbit the galactic center at speeds on the order of a hundred miles every second. If you have a journey that takes a full year, your destination could be three billion miles away from where it was when you started. This also doesn't account for the changing locations of the billions of stars and planets between you and your destination, which might act as road blocks to your journey.

You may think that it would be easier to have a computer guide a ship from the surface of a planet to join a battle in low orbit. In some respects, it is. The journey is much shorter and has fewer obstacles once you get out of the hangar. There are still challenges, though. How does the ship know where the droid control ship is? The ship cannot directly sense it. There aren't mile markers along a highway to indicate where it is. The ship could communicate with allied ships about their location, but how did those ships find their way to the droid control ship?

Locating something in space is not a trivial task. Typically, it takes triangulation. The Naboo army would need to have a network of detectors either in space (preferable) or on the ground looking to the sky. Each detector could detect how far away the droid control ship is from itself. Once there are four points of reference (four detectors have a reading), the droid ship's location can be specified at a particular point in space and time, and a computer can calculate how to move your ship from its current location (relative to those same detectors) to the droid control ship.

THE PHYSICS OF REAL LIFE

How close are we to having guidance computers capable of making these calculations? Well, we developed computers capable of this type of calculation about sixty years ago. Although not for interstellar travel, these computers were used to calculate missions to the Moon, Mars, and every other planet in our solar system (and more!). If we take the Moon as an example, it took about four days, six hours, and forty-five minutes for *Apollo 11* from takeoff to landing. Since the Moon orbits Earth about every twenty-seven days and eight hours, the Moon passed through about 16 percent of its orbit between the takeoff of *Apollo 11* and the landing. With the radius of the Moon's orbit being (on average) about 240,000 miles, that means the Moon traveled about 235,000 miles in the time it took *Apollo 11* to travel the 240,000 miles from Earth to the Moon.

DOING THE NUMBERS

There are a few things to note about the numbers cited for the *Apollo 11* mission. First of all, the fact that the Moon had moved approximately the same distance as it is from Earth is a fluke. It is a consequence of the fact that the time it took *Apollo 11* to reach the Moon is a little under one-sixth of an orbit ($\frac{1}{6.38}$ to be precise). This means the angle that the Moon moved through in this time is close to one radian (or the angle corresponding to traveling one radius-worth of distance around a circle). Also, *Apollo 11* did not actually travel in a straight line from Earth to the Moon. It flew into Earth orbit before boosting to transfer to an orbit of the Moon before landing. The precise distance traveled is difficult to calculate without more data than is publicly available.

THEY DO MOVE

Technically, our destinations are moving, just ever so slightly. Plate tectonics move on Earth's surface about two to four inches per year (on average). Obviously, this is not a concern on a day-to-day basis, but cataclysmic events can cause larger shifts in a short amount of time. For example, after the Fukushima earthquake, the main island of Japan shifted approximately eight feet. Again, on the scale of Earth, eight feet is not much, but that is significant for an isolated event's effects.

In principle, our computers can calculate interstellar travel as long as we know the locations of the star (or nearby stars that may get in the way during our journey). We are still measuring the location of neighboring stars to create a catalogue of their positions. For example, the UCAC4 catalogue compiled by the US Naval Observatory has about 113 million stars in it. That may sound like an impressive number until you realize that it is only 0.113 percent of the estimated total number of stars in the Milky Way.

Closer to home, we use guidance computers constantly during air travel. Airplanes are mostly flown via autopilot, which means turning over the control of the plane to the guidance computer. The location of the airplane can be tracked via GPS very accurately, and (fortunately) the destinations are not moving around like they do in interstellar or interplanetary travel.

HOLOGRAMS

"Help me, Obi-Wan Kenobi. You're my only hope."
—Princess Leia Organa (Episode IV)

WHEN Episode IV, scene when Obi-Wan meets Luke

WHERE Tatooine, Obi-Wan's hut

CHARACTERS Princess Leia Organa, Luke Skywalker, Obi-Wan Kenobi, R2-D2

PHYSICS CONCEPTS Light

SHORT INTRODUCTION/BACKGROUND

Over the past few centuries, the recording and projecting of images has evolved greatly. One technology that has been elusive thus far is the three-dimensional projection of a movie. We have developed ways to make a 2-D picture appear as 3-D (a traditional hologram), but we have not found a way to project a three-dimensional image. How far are we from projecting 3-D images and recordings?

BACKSTORY

Holograms play an important role in communications in the Star Wars galaxy. At the beginning of *A New Hope*, Leia sends a holographic message to Obi-Wan, hoping to recruit him to help defeat the Empire. Part of the reason Luke is so eager to rescue Leia when he discovers she is on the Death Star is that he feels like he knows her already from this hologram. This could also be why the emperor communicates by hologram so often—his physical image makes his tone more threatening. Regardless of the reason, holograms are ubiquitous in Star Wars, but are they even possible?

THE PHYSICS OF STAR WARS

Typically, in order to project an image there needs to be some surface on which to project it. When we watch a movie at the theater, we are seeing a 2-D screen onto which a series of images is projected. How can the holograms in Star Wars be projected into thin air?

There are a few possibilities. First, it is possible that along with the light from the projector, R2-D2 is also emitting a faint amount of smoke to scatter the light in all directions. For us to be able to see something in a particular location, light needs to come from that location and hit our eyes. It is easy for a projector to send light in a straight line (and for that light to be reflected back to our eyes by a screen), but how can R2-D2 cause light to go to a certain location in space and then be scattered or reflected in all directions?

The short answer is, this technology doesn't exist, but we can theorize about how it could be done. If you have ever seen a laser pointer in a dusty or foggy room (think laser tag), you may have noticed that you cannot see the beam unless it is passing through the dust. Much like two billiard balls colliding, some of the light will bounce off the dust and go to the sides. That is why smoke could act as a "screen" for R2, but it doesn't explain the reflection of a coherent 3-D image to viewers on all sides.

If we wanted to produce a 3-D image via the scattering of light, we would need to find a way to scatter different bits of light in different directions from the same location. For instance, we need Leia's face to scatter back toward R2-D2 and we need the buns in her hair to scatter to the sides of her head. On top of that, we need the back of her head to scatter not at all! This is quite an engineering challenge, but there is a way it could be done—although it would distort the image coloring quite a bit.

Different wavelengths (think colors) of light will scatter off particles of different sizes. The sky is blue because shorter wavelengths (blue) scatter better in the air than longer ones (red). When there is a fire, and larger dust and ash particles enter the air, we see more scattering from the reds. So, in theory one could get the reds to scatter more than the blues by emitting a cloud of dust with the hologram. If we projected different colors on different sides of the image, we could approximate this. For instance, if her hair were projected with red light and the back of her head with blue, the dust would scatter the red much better than the blue, allowing her buns to be seen from the side much better than the back of her head.

This does not address focusing the light in a particular location, though. We need the image to form at a particular point in space. To accomplish this, we need to have lenses focus the light. This technology is very old; statistically speaking, you are probably using some form of corrective lens to be able to read these words right now.

From this description, it may sound as if it's possible to produce a hologram with a few colored lasers and a fog machine. It would in fact require microscopically precise engineering to focus various lasers on particular grains of dust to create the appropriate scattering effect. At this time we do not have this technology.

THE PHYSICS OF REAL LIFE

Although we have not been able to construct a true 3-D projection, there have been a number of approximations with varying levels of sophistication. Let's begin with a description of how a traditional hologram works (a 2-D picture that looks like it has depth to it).

In order to make a hologram, a laser beam is split in two; half of it strikes a holographic plate (the equivalent of film in a traditional camera) without any changes, while the other half is reflected off of the object you want to represent and onto the plate. The first beam is used as a reference beam and the second beam interferes with the reference beam in ways that encode the 3-D shape of the object. When you want to display the hologram, you just have to shine a beam that matches the reference beam off of it, and the image will become apparent. Think of it like a simple math equation: reference beam + object beam = interference pattern. If you want a simple picture of the object, just subtract the reference beam from both sides of that equation.

Perhaps the two most famous attempts at producing a 3-D movie hologram were CNN's election night coverage during the 2008 presidential election and Tupac's appearance at Coachella in 2012. Neither of these was a true hologram, though.

In the case of CNN, an array of thirty-five cameras was set up around correspondent Jessica Yellin. These cameras recorded her from all sides and fed those images into a computer. The computer, using the precise locations of the cameras relative to her, was able to construct a 3-D model. The computer then sent that model to CNN headquarters and the image was superimposed on the broadcast signal. Disappointingly, Wolf Blitzer was actually just talking to a spot on the floor rather than a hologram.

The Tupac hologram was both simpler and more rooted in history. Tupac was made to appear onstage via a process known as "Pepper's ghost." Pepper's ghost was first described by sixteenth-century scientist Giambattista della Porta in a work entitled *Magia Naturalis*, or *Natural Magic*. The effect is named for a British scientist, John Henry Pepper, after he popularized the technique in the 1800s. To this day, the technique is still used in places such as Disneyland and Disney World. Anybody who has gone on the Tower of Terror or the Haunted Mansion has seen Pepper's ghost in action.

The effect is possible, roughly speaking, because glass can be both transparent (you see through windows every day) and reflective (sometimes you catch a glimpse of yourself as you look through). As a result, it is possible to reflect light off glass in such a way that it appears the light is coming from behind the glass. In most cases, this is accomplished by having a fully dark room with one well-lit object. The light from that object will reflect off glass between the dark room from the well-lit room, making it look as if the object is hovering in space. Since some light also goes through the glass, the object has an appearance of transparency, much like a ghost.

OTHER TECH

COMMUNICATION

WHEN All films	
WHERE The galaxy	
CHARACTERS All	
PHYSICS CONCEPTS Light, relativity	

SHORT INTRODUCTION/BACKGROUND

Even though it is not often considered explicitly in the films, communication is key to being able to maintain an interstellar society. When the Senate makes the decision to turn power over to an emperor in a time of war, that information needs to be communicated far and wide. In order to coordinate an attack against a planet-destroying base, a fleet needs to be able to communicate with its X-wings. How would interstellar communications work? Could you communicate while in hyperspace? How long would it take for a senatorial decree to reach a place such as Tatooine? Are there any limits to the ways in which we communicate over long distances?

BACKSTORY

From the simple commands between Red Leader and Gold Leader in *A New Hope* to the interstellar issuance of Order 66 from *Revenge of the Sith*, communication is central to coordinating battle. Beyond the battlefield, communication is important to society across the galaxy. Throughout the films we see communication take many forms. Holograms are transmitted via hardware and wireless communications. We see the plans for the Death Star beamed from a satellite dish to the rebel ships. Disrupting the communication between the droid control ship and the droid army successfully ends the Battle of Naboo.

THE PHYSICS OF STAR WARS

When transmitting data, it must be encoded into a signal, which is then transmitted from one point to another. In *The Phantom Menace*, for instance, Qui-Gon sends a blood sample from Anakin back to Obi-Wan for analysis. At first Obi-Wan thinks there is an error in the communication, but after verification, he reads the signal loud and clear (for more on error correction in communication, see Robotic Communication). This exchange tells us a few things about how signals are sent between comlinks: first that there is sometimes interference with the signal, and second, complicated biological information can be sent nearly instantly across a planet.

The first one may seem rather mundane. With every system there are bound to be errors at times. For there to be interference, though, the medium of data transmission has to interact with something in the world around it. If the signal cannot interact, then it cannot be modified. Most likely, these comlinks are sending signals using some form of electromagnetic radiation outside of the visible spectrum (most likely microwaves or radio waves).

Encoding a complicated signal like a blood sample in a signal that can be transmitted nearly immediately is very impressive.

WAVES AND CELL PHONES

Why do most cell phones and Wi-Fi use radio waves and microwaves? It has to do with safety and interference. Any higher energy than visible light is "ionizing radiation," or radiation capable of altering DNA (and thus causing health issues). Sending signals via visible light would be confusing, as there is artificial and natural visible light everywhere that could interfere with the signal. Similarly, all humans (and anything else at room temperature) radiate light in the infrared range, which would also create a lot of interference. This leaves microwaves and radio waves for easy communication.

OTHER TECH

The human genome is about 1.5 gigabytes of data. Including this information along with counts of midi-chlorians and other information could easily exceed 10 gigabytes. Anybody who has tried to download a large file (such as a new operating system) knows that downloading a file of this size usually does not occur immediately. This could indicate that Star Wars technology includes some form of quantum computing, which can store that data significantly more efficiently.

Another good indicator that the characters in the Star Wars saga are using radio waves (probably close to FM) is the satellite dish in *Rogue One* used to transmit the Death Star plans. It's probably more accurate to call it a parabolic antenna, but that's not so important. The important aspect to a parabolic receiver (or transmitter) is that its radius is very close to the wavelength used in its communications. This way, no matter where on the parabolic surface the signal hits, it will be in phase with other pieces of the signal when they reach the receiver (the point located at the center of the parabola). The dish in *Rogue One* appears to be on the order of thirty feet or so. Light that has a wavelength of thirty feet corresponds to an area between AM and FM radio signals.

So, if they're using light to send signals from place to place, how long would it take to send a signal across the galaxy? Well, if the galaxy is indeed 100,000 light-years across, it could take about 50,000 years for light to reach the outer realms. Tatooine won't hear about the emperor taking over for another 49,980 years or so. Yet, Luke knows all about the Empire at the start of *A New Hope*. This must indicate that there is a way to send signals into hyperspace as well. It would be a bit silly if messages were restricted to real space, but people could travel through hyperspace. In that case, the fastest method of communication would be hyperspace envoy.

THE PHYSICS OF REAL LIFE
Radio signals in the US are transmitted in frequency ranges between 87.9 and 107.9 MHz (FM) or about 540 to 1600 kHz (AM).

Other ranges of the radio band are reserved for CB radios, airport control towers, police bands, television stations, etc. Regardless of the range, there are two primary ways to encode the data being sent. These involve modulating the frequency (FM) or modulating the amplitude (AM).

If you consider a wave, it is a series of peaks and troughs that are equally spaced and have the same height. Amplitude modulated signals (AM) will encode information via changing the height of the waves. As an example of a way to do this, you can have the amplitude of the wave change between a height of 1 or height of 2. When the wave has a height of 1, we could call that as a 0, and when the wave has a height of 2, we count that as a 1 (or vice versa). The receiver then receives a signal, which will come in as a series of waves that need to have their height measured. Based off of the series of peaks received, the computer can translate the height of the wave into a binary signal to be processed.

In the case of an FM signal, the frequency of the wave is modulated. The receiver can decode it by considering how frequently the peaks reach the receiver. For example, suppose the receiver is looking in intervals of, say, one second. If the receiver receives five waves in one second, we can consider that a 0 and if the receiver receives ten waves in one second, then we count that as a 1. These signals can then be translated into binary for the computer to process.

Inevitably, our signals will encounter interference. How does a computer interpret a signal which does not match precisely what we expect (for instance, an AM signal with a height of 1.5)? In communication, redundant information is often included so that even if one bit is messed up, the full signal can be understood by the receiver. There are different ways to check for error detection, but one includes ensuring that your encoding always has an even number of 1s. For instance, if you wanted to send the signal yes (1) or no (0), you would add additional bits so that your message would always have an even number of 1s (11 for yes or 00 for no).

OTHER TECH

AN EVEN ZERO

There is some debate in mathematics of whether zero can be counted as even. For the purposes of computer science, zero is even parity.

For more complicated signals, there can be a convention of groups of bits always having an even number of 1s. For instance, take a signal of n bits (for the sake of specificity, we will use n = 5). We say that the group of bits 1, 3, and 5 need to have even parity, the bits 2 and 3 need to have even parity, and the bits 4 and 5 need to have even parity. How did we determine those groups? We will need to write the bit numbers in binary (001, 010, 011, 100, 101). If you notice, all the numbers with a 1 as the final digit are in a group (1, 3, 5). All the bits with a 1 as the second digit are in one group (2, 3). And lastly, all the numbers with a 1 as the first digit are in a group (4, 5). With this scheme, we can send four signals that can be error checked. For example, let's say we're sending four different signals (say, no, don't know, maybe, and yes). We can say that "no" is encoded as 00000, "yes" is 01111, 10011 is "don't know," and 11100 is "maybe." This way, even if a single bit becomes distorted, we can still figure out what the initial message trying to be communicated is. If you change a single bit of any of those messages, it will still be closer to the correct message than any of the other four.

If the receiver receives another signal, it can probably figure out what was meant to be sent. For instance, if the receiver receives the signal 10111, the parity checks will go as follows. As described previously, we will need to look at the bits in different groups. According to the encoding scheme we are using, each group should have an even number of 1s in it. First, we check the 1, 3, and 5 bits and find that they are all 1s. Three 1s mean one of those bits is an error. Which one? Well, we check bits 2 and 3 and find that they do not have an even number of 1s either. This means that the third bit is an error bit (because it was in both groups with an error), and we can correct it to be a 0 and correctly receive an answer of "don't know."

AUGMENTED REALITY

"His computer's off. Luke, you switched off your targeting computer. What's wrong?"
—Rebel base controller (Episode IV)

WHEN Episode IV, Battle of Yavin	
WHERE Death Star trench	
CHARACTERS Luke Skywalker	
PHYSICS CONCEPTS Community detection, reflection	

SHORT INTRODUCTION/BACKGROUND

As we go through life, we rely on our senses to inform us about the world around us. Unfortunately, our senses are limited. We can only see wavelengths of light between roughly 400 and 750 nm, and we can only hear sounds between 20 and 20,000 Hz. Our brains' storage and recall abilities can sometimes feel very limiting. We have computers that are getting smaller and smaller and better at enhancing humans' capabilities, but very few have those enhancements built into their bodies. How can we enhance our daily lives by augmenting reality? What are the current applications of such technology?

BACKSTORY

During the Battle of Yavin, Luke is meant to use his X-wing's targeting computer to guide proton torpedoes into the Death Star's exhaust port. The computer shows a schematic of the trench and the approach to the target, along with a countdown of the distance to the exhaust port. This is one of many examples of augmented reality in action in the Star Wars films, but it is not clear just how advanced or widely used this technology is. In this instance, Luke opts to use the Force, which succeeds where a previous pilot and targeting computer had failed.

THE PHYSICS OF STAR WARS

Not all augmented reality shown in Star Wars looks especially advanced, but it does share a basic functionality—providing information to the user that is not readily available via natural senses. How might these tools work? In the case of the X-wings' targeting computers, the basic visuals and distance tracking information could be provided in several ways. The Rebel Alliance has a schematic of the Death Star, but that doesn't help with live information (incoming TIE fighters, updating distance to the exhaust port, etc.). For the live data, something must be tracking the Death Star and nearby ships.

The distance to the exhaust port could use a location tracker to compare an X-wing's current location to the schematic. On the surface this seems fairly easy; it happens every time we use Google Maps. In this case, however, even if you have a perfect map of the Death Star, how do you track the location of your ship relative to the map? With GPS we have an array of satellites that communicate their exact positions relative to you. Presumably the rebels did not install satellites orbiting the Death Star, but they could have a similar setup using their larger ships. Perhaps along with providing orders to the fighters, the ships are also tracking their own locations relative to the Death Star and the X-wings. This seems like a roundabout way to accomplish the goal, and it would also mean that if a fighter ship lost signal, the torpedo would be launched at a completely incorrect time.

Another method involves sensing where an X-wing is relative to other objects using something analogous to radar. Radar, an acronym for radio detection and ranging, emits radio waves and measures how long it takes for them to reflect back. Based off this, you can determine how far away things are. Since we know the speed of light (radio waves are a type of light), the time it takes for waves to go from point A to point B and back times the speed of light is the distance between points A and B. For example, if it takes radio waves ten nanoseconds to return, the object reflecting the waves is three miles away. Bats and dolphins do the same sort of thing to find food or predators around them.

Each X-wing could be equipped with radar, as could the larger command ships in the fleet. That way, someone has a complete picture of the operation to help better instruct squadron leaders, but individual ships wouldn't be completely blind if other ships are taken out.

THE PHYSICS OF REAL LIFE

Augmented reality is a relatively old concept (one could argue that radar existed in the animal kingdom well before humans developed it between the late 1800s to early 1900s), but it has become more mainstream recently. One of the first written examples of what we think of as augmented reality came in 1901 in a novel called *The Master Key*, written by L. Frank Baum (author of *The Wonderful Wizard of Oz*). In this book, the protagonist is provided a "character marker," eyeglasses that superimpose a letter on others' foreheads to indicate whether they are cruel or kind, good or evil, etc. This might sound like science fiction, but this type of technology could be in development. With Google Glass, Twitter follower numbers could be superimposed onto people's faces. Admittedly, there may be little correlation between Twitter followers and one's character, but as more and more data is accumulated from different sources, a more accurate algorithm is feasible. The Internet of Things revolution could mean adding to the available data how many times a day you drive your car, take a shower, visit various websites, etc.

The most popular version of augmented reality in 2016 was Pokémon Go. This is a smartphone game that superimposes a cartoon Pokémon onto an image of your surroundings using the phone's camera. This may seem like a silly application of a great technology, but the applications of this are endless. For instance, what if you could walk down the street and look through your smartphone and see the names of restaurants or stores on the street with their Yelp ratings, ratio of CEO pay to median employee pay, etc. What if this didn't even require a phone and was just a function you could turn on from an eye implant? At least one patent application for "smart" contact lenses has been filed (by Google).

Currently, one of the biggest limitations to full augmented reality is the lack of effective image recognition algorithms. Currently image recognition algorithms are in their infancy. This is true both in that it is a relatively new technology, and in that literally the human eye of an infant can do better than most algorithms.

One proposed method of image recognition is a process derived from graph theory called community detection. This involves looking at how data is connected. This technique has been used to find patterns in everything from the atomic structure of amorphous solids to ecological food webs, but it is now finding applications in image analysis. There are many methods you can use, but the ultimate goal is to find data points in different groups. To give a specific example, imagine your friend network on a social media site. A community detection algorithm would be able to find your high school friends versus your college friends versus your family members, etc., based off of mutual connections among your friends.

For the purposes of image analysis, the data points are pixels rather than people. Pixels might be "connected" based on similarity in RGB values (the amount of red, green, and blue needed to make the color of that pixel). The community detection algorithm will group neighboring pixels if they are roughly the same color. This is a rudimentary way to separate, for instance, a hat from the head wearing it in a photo. The pixels of the hat will be in a different community from the pixels of the head because they are (almost always) different colors.

TRACTOR BEAMS

WHEN Episode IV, *Millennium Falcon*'s arrival at Alderaan	

WHEN Episode IV, *Millennium Falcon*'s arrival at Alderaan

WHERE Location formerly known as Alderaan

CHARACTERS Luke Skywalker, Obi-Wan Kenobi, Han Solo, Chewbacca

PHYSICS CONCEPTS Force, electromagnetism, fluid dynamics, pressure

SHORT INTRODUCTION/BACKGROUND

The idea of a tractor beam is reasonably easy to imagine. A battle station or ship shoots out some kind of force that pulls the target closer. Once the object has been caught in a tractor beam, it is difficult to escape its pull. What type of force does a tractor beam use? What is its range? How can you escape a tractor beam if it catches you? Are there any unintended consequences of using a tractor beam? Could a tractor beam capture more than one ship at a time?

BACKSTORY

Upon arrival at Alderaan (or what's left of it), the *Millennium Falcon* and its crew are captured by a tractor beam from the Death Star. Trapped in the ray, the ship is slowly brought into the hangar bay without any chance for escape. We see tractor beams used again in *The Force Awakens* (again on the *Millennium Falcon*) when Rey and Finn, escaping from Jakku, are captured by Han and Chewie. In both cases, thrust from the ship seemed to be disabled and it's drawn into the other ship.

THE PHYSICS OF STAR WARS

There are a few properties that seem to be consistent among the tractor beams shown in the movies. The beam creates an attractive force between two objects, and the target object does not seem capable of escaping. We will take these two properties one by one.

First of all, how could you create a "beam" of attractive force? The force would either have to be gravitational or electromagnetic in origin as the other two fundamental forces (strong and weak nuclear) have a range on the atomic scale. To create a gravitational attractive force, you would have to be able to manipulate space-time around a target. You could do this such that the object is suddenly trapped inside a gravitational well (think of it as being at the bottom of a very steep invisible pit). No matter how much thrust the ship applied, it would not be able to escape the pull. You could then manipulate that well so that the object would slowly approach you.

This sounds like a very clean way to accomplish the goal. There is only one problem: we don't have the slightest ability to do this.

Could the characters in Star Wars have developed such technology? Perhaps, but there are some other issues associated with this solution. In order to warp space-time, you need high concentrations of mass or energy. Since the tractor beam is clearly not emitting mass, it must be sending out gravitational energy. If the tractor beam was too focused, it could create a black hole, which would devour your target (and potentially you).

If you were to try to accomplish this goal via electromagnetism, you could try in one of two ways. First, and perhaps most obviously, you could just use a giant magnet. This could work very well for what is shown, but it could also be ineffective. In order to evade it, all you would have to do is make your ship out of something that is

---

KUGELBLITZES

I would be remiss as a physicist if I didn't take a moment to let you know that a black hole created by a high density of electromagnetic radiation is known as a kugelblitz. They are only theoretical, but with a name like kugelblitz, I'd like to think we'll verify their existence someday.

not magnetic (aluminium, for example) and you'd never be able to be captured by a tractor beam! You can also focus electromagnetic energy to the point that it warps space-time enough to accomplish the gravitational pull. However, this has the same problem as using any kind of energy to warp space-time.

How escapable are these different kinds of tractor beams? If you were to warp space-time around the ship, you could effectively prevent it from escaping, no matter how much thrust it used. If you were to hold up a giant magnet, though, all the ship would need to do is exert thrust greater than your magnetic force. One issue with both of these methods is that (in principle) the ship could come at you faster than you would like (unless you had some other method of disabling its thrusters). If you are pulling the ship toward you, you are creating a path with lowering potential energy. In that situation, the ship could turn on you and charge. It would most likely be a kamikaze mission, but if you needed to make sure the crew was taken alive, you might be in trouble.

THE PHYSICS OF REAL LIFE

You may have read that tractor beams actually exist. If your goal is to drag microscopic particles over distances of centimeters, then you are in luck. Scientists have created both sound-based and light-based tractor beams. The most effective ones (and ones able to work on larger particles) are the sonic tractor beams. It should

---

BERNOULLI'S PRINCIPLE

Bernoulli's principle states that moving air is at lower pressure than stationary air. If you suspend a Ping-Pong ball above a hairdryer, it will stay in the column of moving air because as it moves from side to side, it encounters higher air pressure that pushes it back into the column of air. Similar processes happen with sonic tractor beams.

be noted before I go any further, though, that development of a sonic tractor beam would be worthless in space where there is no atmosphere through which the sound can travel (see the Seismic Charges section for more on that).

The idea behind a sonic tractor beam is that an array of speakers send out sound waves, which interfere with each other to create pockets of high pressure and low pressure in the air. Those pockets can be manipulated to exert control over a particle (often a drop of colored liquid is used for demonstration purposes). This process is not dissimilar to a more controlled version of holding a Ping-Pong ball in place using a hairdryer.

The contours of the sound waves controlling the particle can be made into different shapes depending on the type of control you need. For instance, they can be in the form of two beams on either side to act as tweezers, which can move closer together and farther apart. This allows you to compress the particle or give it more range of motion as you bring it closer to you. You can also use these "tweezers" to rotate the particle in the air. This could be useful if you need to make sure that the ship you're targeting faces away from you (so can't charge you).

Although these beams are only able to manipulate a dust particle (or something slightly bigger) over short distances, this concept could be expanded to affect larger objects. The primary issues are that larger objects need larger forces to manipulate them. Larger forces arise from larger pressure gradients. For instance, if you want to exert a force equivalent to gravity on a person, you would need to reduce the air pressure to 83 percent of atmospheric pressure at the feet (while maintaining normal air pressure at the head). This would be approximately equivalent to creating the atmospheric density of Denver at a person's feet while maintaining sea-level density at his head.

This sounds like no big deal, right? Well, there is an issue there. To create a disturbance in air pressure this large requires very loud sound. We could have it be in ranges outside of human hearing, but that doesn't fix the primary issue with how loud these sounds need to be. A pressure gradient this large (about 20kPa over the height of a person) would require sounds on the order of the loudness of the Krakatoa eruption (see Seismic Charges for why this is bad). Even if this is outside of the realm of human hearing, it would require energy well beyond what it's worth (and that's just to move a human, not a full spaceship).

APPENDIX A.
GLOSSARY OF STAR WARS TERMS

Alderaan
A terrestrial planet that was home to Princess Leia Organa. It was destroyed by the first Death Star.

Baradium
A fictional element that is highly volatile. One location where baradium is mined is on the planet Denova.

Bespin
A gas giant that is a major source of tibanna gas. It is home to Cloud City.

Bongo
A Gungan submarine.

Booma
A Gungan grenade-style weapon.

Boonta Eve Classic
The annual podrace on Tatooine held to celebrate the holiday Boonta Eve.

Bothans
An alien species credited with stealing the plans for the Death Star II. They are a short humanoid species covered in fur.

Carbonite
A fictional material that is used to store tibanna gas for transport. It is made through a process which involves converting carbon gas into a liquid form. It was notably used to freeze Han Solo in *The Empire Strikes Back* (Episode V).

Cato Neimoidia
A planet colonized by the Neimoidian. It is home to the headquarters of the Trade Federation.

Clone Wars
A three-year conflict between the Galactic Republic and the Separatists. This was engineered by Darth Sidious to eliminate the Jedi order. This is not to be confused with the title of the TV series, *The Clone Wars*, which chronicles the events of the war.

Collapsium
A fictional element that implodes when ignited.

Coruscant
A planet city located in the galactic core that is the governing center of the galaxy.

Dagobah
A swamp planet that Yoda calls home.

Death Star

A moon-sized battle station capable of destroying a planet in a single blast. The original incarnation of this station was destroyed in the Battle of Yavin. After partial reconstruction the second incarnation was subsequently destroyed in the Battle of Endor.

Endor

A forest moon that is home to the Ewok race. It was the location of the Imperial shield generator for the Death Star II and site of the Battle of Endor, in which the Ewoks joined with the Rebel Alliance against the Imperial Army.

First Order

A faction that arose out of the collapse of the Galactic Empire. They constructed Starkiller Base. Notable members include Kylo Ren, Supreme Leader Snoke, and Captain Phasma.

The Force

An energy field that permeates and connects all living things in the galaxy. Individuals who are sensitive to the Force can tap into it, and they can enhance their capabilities through proper training.

Galactic core

A region of the galaxy near the center. Examples of planets in the galactic core include Coruscant and Hosnian Prime.

Galactic Empire

A constitutional monarchy that arose when then-chancellor Palpatine consolidated his power through the mostly manufactured threats to the Republic following the Clone Wars.

The galaxy

The setting for most of the Star Wars films. It is spiral in shape and has about 100 billion stars orbiting a galactic center. There are different regions of the galaxy, including the galactic core, the mid rim, and the outer rim.

Gungans

An amphibious species native to Naboo. Notable examples of Naboo characters include Boss Nass and Jar Jar Binks.

Hoth

The sixth planet in the Hoth system. It is an icy planet and home to a rebel base at the start of *The Empire Strikes Back* (Episode V). It also acts as the setting for the Battle of Hoth.

Hyperdrive

A propulsion system used to access hyperspace.

Hyperspace

Extradimensional space accessible by traveling faster than light.

Inner rim

A region of the galaxy just outside of the densest part. It is home to many planets, most notably Jakku.

Invisible Hand

A ship commanded by General Grievous.

Jedha

A desert moon of the planet NaJedha that is home to a major mine for kyber crystals.

Jedi

An order of Force-sensitive individuals who primarily represent the light side of the Force. The preferred weapon of a Jedi is the lightsaber (often blue or green in color). Their primary adversaries are the Sith.

Kamino

An ocean planet that was home to the facilities responsible for creating and training the clone army.

Kyber crystal

A Force-attuned crystal used to construct lightsabers, among other uses.

Lightsaber

A sword-style weapon whose blade is made up of either plasma or light (depending on the source). It is the primary weapon of Force-sensitive individuals.

MagnaGuard

Elite battle droids favored by General Grievous.

MC75 star cruiser

A class of star cruiser used primarily by the Rebel Alliance.

Midi-chlorian

A sentient microscopic organism. Force-sensitive individuals have a high midi-chlorian count.

Mid rim

A region of the galaxy between the galactic core and the outer rim. Examples of planets in the mid rim are Naboo and Kashyyyk.

Millennium Falcon

A ship used primarily by Han Solo and Chewbacca. Although externally it appears to be a run-down light freighter, its internal systems were modified extensively for more effective smuggling.

Mos Eisley

A large spaceport on the planet Tatooine.

Mustafar

A small, highly volcanic planet in the outer rim of the galaxy. It was the location of the lightsaber duel between Obi-Wan Kenobi and Anakin Skywalker in *Revenge of the Sith* (Episode III). It later became the location of Darth Vader's castle.

Naboo

A pastoral planet that is home to the Gungans, Padmé Amidala, and Sheev Palpatine.

NiJedha

An ancient city on Jedha that was home to the Temple of the Kyber, which was looted of its crystals to help construct the Death Star.

Outer rim

A region of the Galaxy far from the galactic center. Many planets in the outer rim serve as hideouts or homes for outlaws due to their distance from the center of governmental power. Planets in the outer rim include Tatooine and Dagobah.

Padawan

A Jedi in training.

Polis Massa

An asteroid where Anakin and Leia were born.

Rishi Maze
A dwarf galaxy orbiting the galaxy. It is used as reference for the location of the planet Kamino.

Scarif
A planet in the outer rim that was the location of the Imperial security complex, where the plans for the Death Star were kept. It was the setting for the Battle of Scarif between the Rebel Alliance and the Imperial Army, which ultimately led to the rebels obtaining the plans.

Sith
An order of Force-sensitive individuals who primarily represent the dark side of the Force. The preferred weapon of a Sith is the lightsaber (typically red in color). Their primary adversaries are the Jedi.

Star Destroyer
A kite-shaped ship used by the Galactic Empire.

Starkiller Base
A planet that was modified to act as a massive weapon capable of destroying entire star systems.

Tatooine
A desert planet located in the outer rim. It was home to Luke Skywalker, Jabba the Hutt, and Anakin Skywalker.

Tibanna
A fictional element used in a variety of antigravity weapons including blasters and repulsorlifts. The planet Bespin is a primary source of tibanna gas.

TIE fighter

A fighter ship with twin ion engines used primarily by the Imperial fleet for close-range fighting.

Toydarian

A winged species native to Toydaria. They are immune to Jedi mind tricks. A notable example is Watto, owner of Anakin and Shmi Skywalker.

Trade Federation

The interstellar conglomerate that manages much of the trade throughout the galaxy.

X-wing

A ship used primarily by the Rebel Alliance for close-range fighting.

Yavin 4

A moon orbiting Yavin that was the setting for the destruction of the first Death Star.

APPENDIX B.
GLOSSARY OF PHYSICS TERMS

Absolute zero
The coldest possible temperature (about $-460°F$), at which all motion ceases.

Alpha Centauri
The closest star system to the Sun. It consists of several stars including the red dwarf Proxima Centauri.

Antimatter
Material composed of particles that annihilate when they come into contact with their partner, matter.

Big bang theory
A theoretical explanation for the beginning of the universe that describes a stretching of space-time from a hot, dense mess of particles to the universe we see today.

Binary star system
A pair of stars that co-orbit each other. Sometimes planets orbit binary stars. They are relatively unstable compared to their single-star counterparts.

Black hole
The densest object in the known universe. Black holes distort space-time as well as the path of light.

Capacitor
An electrical component capable of storing energy that can be released quickly.

Casimir force
Attraction between two metal plates when placed close together. It is an inherently quantum mechanical force.

Dark energy
Theoretic energy of space-time used to explain the expansion of the universe.

Dark matter
Theoretic material that is proposed to describe a number of anomalies in astronomical measurements.

Dwarf galaxy
A collection of stars approximately one thousand times smaller than a full galaxy.

Electrode
Part of a circuit connecting a conductor to a nonmetallic part (e.g., semiconductor).

Electromagnetic field
A vector field which surrounds a charged object which can exert forces on other charged particles.

Endothermic reaction
A chemical reaction in which energy is absorbed.

Exothermic reaction
A chemical reaction in which energy is released.

Extrasolar planets (*also* exoplanets)
Planets that exist outside of our solar system.

Force
A push or a pull.

Galaxy
A collection of stars numbering on the order of 100 billion to 1 trillion. A galaxy comes in one of many shapes including spiral and elliptical.

Galileo
A probe sent to study Jupiter and its moons.

Gravity
One of the four fundamental forces of the universe. It is responsible for holding us on the surface of the Earth, the orbiting of planets around stars, and stars around the supermassive black hole at the center of a galaxy. It is the weakest of all fundamental forces. The most precise description of gravity is general relativity.

Hawking radiation
Energy emitted by black holes that ultimately leads to their evaporation.

Hubble telescope
An instrument used to image distant parts of the universe.

International Space Station (ISS)
A habitable satellite in low-Earth orbit that launched in 1998. It is used for microgravity experiments.

Kepler telescope
An instrument used to detect extrasolar planets.

Length contraction
Shortening of length measurements due to relativistic effects.

Lepton
A class of subatomic particle that includes electrons.

Light
An electromagnetic wave that has different properties depending on its wavelength. It is visible to humans in a limited range (~400–750 nm).

Light-year
The distance that light travels in one year, or about 6 trillion miles.

Microbiome
The microorganisms of a particular environment (e.g., inside the human body).

Milky Way
A spiral galaxy where Earth is located.

Momentum
Conserved quantity that increases with speed and inertia.

North Pacific Gyre
A region in the North Pacific Ocean where currents go clockwise around a high-pressure point near Hawaii.

Orbit
The circular path that a body follows as it travels around another object.

Phase diagram
A chart showing the stability of different states of matter (solid, liquid, gas) as parameters such as pressure and temperature change.

Photon
A particle of light.

Planet
An astronomical body orbiting a star that is (1) massive enough to be rounded by its own gravity, (2) not massive enough to initiate fusion, and (3) has cleared its region of planetesimals.

Plasma
A state of matter in which the temperature is high enough that electrons and nuclei cannot stay bound. It is like a soup of charged particles.

Space-time
Shorthand for the three spatial dimensions and time.

Supercluster
A large collection of thousands of galaxies.

Superconductor
Material capable of conducting electricity with zero resistance.

Supernova
An explosive death of a massive star.

Tachyon
A theoretical particle capable of moving backward in time.

Tesseract
A four-dimensional analog of a cube.

Tidal force

The difference in the strength of a gravitational force on one side of an object compared to the other. For instance, the gravitational pull from the Moon on Earth is stronger on the side closer to the Moon.

Time dilation

Lengthening of time measurements due to relativistic effects.

Universe, observable

Everything that can currently be detected from Earth.

Vacuum

Space void of matter.

Van der Waals force

A weak attraction between molecules caused by polarization of charge inside the molecules.

Wave

Any oscillation that travels through space and time. It can be the oscillation of an electromagnetic field (light), water, or a medium in the form of sound.

Wormhole

A theoretical connection between two distant points in space using additional dimensions.

INDEX

INDEX

ABOUT THE AUTHOR

Patrick Johnson, who holds a PhD in physics from Washington University in St. Louis, fell in love with Star Wars and science at a very young age. Around three years old, he was heard saying, "When I grow up, I want to be a scientist with dirty hair." A few years later, his father, Eric, showed the original Star Wars film to him. Being fascinated by the idea of a laser sword, Patrick immediately went on to watch the next two films, after which, he decided that he wanted to meet an Ewok.

As Patrick went through college at the University of Dayton, he began TAing for physics classes, but his favorite was a class for non-scientists. Through that class, Patrick found his love for science communication. It was also during this time that he began to explore different ways to connect physics to people's lives.

These two loves came together for the first time when Patrick asked a question on an exam based off of Yoda using the Force. Star Wars references continued to show up in classes and eventually led to a series of videos on the physics of Star Wars for Georgetown University in DC, his current employer. This book represents the union of two of his biggest passions.